KU-532-585

Geographies of Health

For Caroline and Anna

Geographies of Health:
An Introduction

Anthony C. Gatrell

BLACKWELL
Publishers

The moral right of Anthony C. Gatrell to be identified as author of this work has been asserted in accordance with the Copyright, Designs and Patents Act 1988.

First published 2002

2 4 6 8 10 9 7 5 3 1

Blackwell Publishers Ltd
108 Cowley Road
Oxford OX4 1JF
UK

Blackwell Publishers Inc.
350 Main Street
Malden, Massachusetts 02148
USA

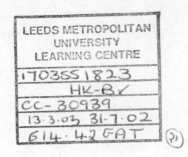
British Library Cataloguing in Publication Data

A CIP catalogue record for this book is available from the British Library.

Library of Congress Cataloging-in-Publication Data

Gatrell, Anthony C.
    Geographies of health: an introduction / Anthony C. Gatrell.
        p.   cm.
Includes index.
    ISBN 0–631–21984–6 (hardback : alk. paper)—ISBN 0–631–21985–4 (pbk. : alk. paper)
        1. Medical geography.   2. Environmental health.   I. Title.
    RA792 .G38 2002
    614.4'2—dc21

                                                                    2001000217

Typeset in 10.5 on 12 pt Sabon
by Best-set Typesetter Ltd., Hong Kong
Printed in Great Britain by MPG Books Ltd, Bodmin, Cornwall

This book is printed on acid-free paper.

CONTENTS

LIST OF FIGURES

LIST OF TABLES

PREFACE

Fifteen years ago, when I first became seriously interested in geographical studies of disease and health service provision, I encountered expressions of surprise among colleagues in the medical and allied professions. In the last four or five years, however, the fact that a geographer is working in this field occasions less curiosity. I can only think that this is due in part to the success that geographers have had in publishing in health-related and medical journals and in working collaboratively with health professionals; but also to the fact that place, space, and environment – long-standing concerns of the geographer – have permeated much of the health and medical literature, as well as the public consciousness. As I begin to show in the first chapter, where you live has all sorts of impacts on health and health care delivery, a   as a whole is an attempt to develop this theme.

The book begins with two chapte   her, lay out some basic principles and concepts and illustrate the   approaches to the geography of health. It is this diversity that lets n   out "geographies" of health; in other words, the richness of variety   ves that may be drawn upon to conduct investigations into place-t   ion in health and health care. The first chapter describes case st   sent different styles of inquiry. The second explains in more de   these styles are, and uses other studies to illustrate them. Toget   rs give a "map" of the variety of approaches to the subject and w   timulate the reader to reflect upon the merits and otherwise of co   proaches. The third chapter introduces readers to a variety of m   t geographers use in the analysis of health data.

The material that follows is di   ed into two broad sections. The first of these comprises three chapters that explore three core themes in the social geography of health: health inequalities; the provision and use of health care services; and migration and health. I examine, at a variety of geographical scales, the patterning of inequality of health outcomes; to what extent are there place-to-place differences in mortality and illness, and what accounts for these differences? To what extent are there inequalities, expressed geographically, in the supply and uptake of health care services, whether at the level of primary care or in access to advanced medical interventions? Last, what is the association between population movement and health? As we shall see, there is a variety of ways in which migration impacts on both health outcome and on health care delivery, as well as *vice versa*.

In the third section of the book I turn attention more to the "environmental" than the "social," looking at the range of associations between environmental quality and health. This is examined in three separate chapters. First, I look at the evidence implicating air quality and pollution in the geographical patterning of illness. Next, I examine associations between water quality and pollution and health outcome. Last, I consider the important question of links between global environmental change and human health. But I hope it becomes clear that separating the "environmental" from the "social" is not especially productive.

The book derives in part from a course that I have taught at Lancaster University, to students in their second and third (final) year of undergraduate study. I hope that the book will be of use to those students beginning research in this field, as well as to undergraduates studying other subjects, and in other countries. I have endeavored to cast my net widely for examples and illustrations. Inevitably, since I am trying to synthesize a substantial literature, much of which has been published by British and American researchers, many of these examples are drawn from Britain and the USA; but not exclusively so. If the book encourages people there and elsewhere to pursue similar and new directions of inquiry, I shall be well pleased. Yet a text like this can only be a partial account, reflecting the interests and expertise of its author. I hope that, in covering aspects of both the social and physical environment, I have touched on a sufficiently wide range of material to ensure that the book is useful as a course text. But for some colleagues and students there will be omissions. For example, while I have plenty to say throughout about health status in the developed world there is much more to be said about the developing world. Equally, I am conscious that mental health and disability are under-represented in the book.

My geographical background is, I imagine, easy to detect in what follows, and I hope that the book will find a readership among fellow geographers. But I very much hope that it will interest an audience that does not have any formal geographical training: those in the health and medical professions. The boundaries between "health geography," epidemiology, public health, and social policy are indeed fluid, as instanced by the considerable body of material, examined later, that is taken from epidemiological, public health, and related journals. Anyone who wishes to travel further along this road will have to acquaint themselves with several sections or floors of a good library (if not several libraries)! I can promise that this trip will be a rewarding one.

For those who are curious, it may be of interest to know what prompted my initial interest in the subject of the geography of health. One starting point was the concern, expressed in the media in the early 1980s, about the safety of the whooping cough (pertussis) vaccine in the UK; I obtained some small-area data on variations in uptake of childhood immunization and sought to use Population Census data to model this variation. Another motivation was publicity given to apparent "clusters" of children born with eye malformations in the vicinity of a high-temperature incinerator in South Wales; since this purported to highlight a "spatial" as well as an environmental issue, I thought it was worth exploring. In the event, after several years trying to establish a comprehensive database with a Consultant Ophthalmic Surgeon, Richard Collin, a national study was undertaken by a group given the resources to explore it thoroughly. But these early interests led to other collaborative work, both on specific diseases and, more broadly, on health inequalities. The latter has meant that an engagement with non-quantitative social scientists has paralleled my continuing interests in quantitative spatial analysis. Much of the satisfaction I gain from my work derives from working happily with qualitatively minded sociologists on one project and statisticians on others!

In 1996 Lancaster University created an Institute for Health Research, following substantial investment in a research program by the North West Regional Office of the NHS Executive. I want to record my thanks for the support and encouragement offered by several individuals during my period as Director of the Institute. Maggie Pearson, Director of Research & Development in the Regional Office, herself a geographer by background, has been an influential figure, while more locally Lois Willis, Chief Executive of Morecambe Bay Health Authority, has been a source of encouragement. The Institute was given an early impetus by a number of senior figures at Lancaster University, notably Nick Abercrombie and John Urry. Nick has continued to offer outstanding support, as too has Bob McKinlay. I have also been privileged to work with a number of other colleagues at Lancaster University, geographers and non-geographers alike, who have offered intellectual and social support; particular thanks are due to Sally Hollis, Diane Langley, Christine Milligan, Maggie Mort, Colin Pooley, Carol Thomas, and John Welshman among the academic staff. Elsewhere in the Institute, Pat Clelland and Hayley Pinington have proved wonderful colleagues to have around. I owe a considerable debt to Chris Beacock, Staff Cartographer in the Geography Department at Lancaster University, for his skill in producing the maps and diagrams.

I have also had the pleasure (would that they might say the same!) of supervising, or helping to supervise, a series of research students in Geography, some of whom have now become great friends. I would therefore like to thank Paul Boyle, Jo Briggs, Jan Hopkins, Yo-Eun Kim, Simon Kingham, Len Mussell, Jan Rigby, and Clive Sabel for teaching me as much as I taught them. Jan Rigby, herself now a much-valued colleague at Lancaster, has read and commented in detail on the text and I am indebted to her for helping me re-think some of what I have tried to do here.

The international community of health geographers is still relatively modest in size, though expanding gradually. I have benefited enormously from interacting with many individuals, particularly at the series of biennial international symposia that rotate among Britain, Canada, and the USA. The work of this group is, hopefully, reflected adequately in what follows and it would be invidious to name only a few individuals when so many deserve credit for contributing indirectly to this book. The suggestions for further reading, and the list of references, indicate where my debts lie.

My mentor and former supervisor, Peter Gould, died as this book was nearing completion and I pay tribute to him for his fine geographical scholarship and friendship over many years. I am saddened that he did not get to read the book – though he did comment on its proposed contents – but I do know that he would have approved of much of its substance. I sincerely hope that I have been able to impart to my audiences something of the geographical imagination that he always transmitted elegantly in print and in public presentations. He is much missed.

Outside the academic world, I continue to owe debts to my parents and brother for their encouragement, while, elsewhere, Pam and Mike Nicholson, and Max Simon, have taken a keen interest in this project. I am grateful for this. I have been fortunate to have help in all sorts of ways from my wife Caroline, whose background in health services management and expertise in numerous other areas have proved invaluable. She has helped me focus my thinking on many occasions. Our daughter, Anna, was born as the idea for this book took shape. Fortunately, the book has not been so long in the making that she can read it yet! But if and when she does I hope she will know, as her mother does already, that the dedication is heart-felt.

Tony Gatrell
Lancaster
2000

PART I

DESCRIBING AND EXPLAINING HEALTH IN GEOGRAPHICAL SETTINGS

CHAPTER 1

INTRODUCING GEOGRAPHIES
OF HEALTH

It is difficult to pick up a newspaper or magazine, or to listen to a television or radio bulletin without being alerted to a health "problem," whether this be concerned with poor quality screening for a disease, a possible risk arising from environmental contamination, or the lack of access to the basic resources needed to live a healthy life. We may find it hard to define what we mean when we say we are healthy – though perhaps easier to describe our feelings of being unwell – but "health" is something we all have, or have had. At the same time, we all live somewhere on the earth's surface; most of us have homes, some have workplaces, others spend time at school or engaged in retirement activities. We occupy locations and, in the course of our lives, move from place to place. We all have our own "geographies" as well as our biographies.

I aim to convince you in the course of this book that our "health" and our "geographies" are inextricably linked. The screening we get for diseases will be available differentially from country to country or from one health region to another. Where you live affects the treatment you get. The risks arising from environmental contamination, be this poor air quality or polluted groundwater, are not uniform over space. If you live on a busy main road, very close to a source of electromagnetic radiation, or near a site disposing of hazardous waste, you may be more at risk of illness than others who do not. Where you live affects your risk of disease or ill-health. Access to basic resources, such as good, affordable food, clean water, decent housing, and rewarding (and properly rewarded) employment is also geographically differentiated. Where you live affects how accessible or available are such resources.

If you approach this book with a geographical imagination already well developed via other books or courses you will, I hope, find the statements above uncontroversial, though if you are new to the geography of health I intend to persuade you that the subject of "health" is a rich source of material that bears study by the geographer. If I can stimulate more of you to take up the "geography of health" as an area for further enquiry or research I shall be well pleased! But I hope too to interest other readers; those who come to the subject with a background in health research – perhaps public health, general practice, or nursing – or another social or environmental science, and who are intrigued by what a geographer might have to say on "health."

To set the scene, I need to explain some basic ideas and concepts. For some readers (perhaps some geographers) I need to say something about health, illness, disease, and disability. These are high-level concepts which are far from unproblematic, but I will endeavour to say something about these in this first chapter so that some of the early material makes some sense. For other readers (primarily non-geographers) I need to introduce some fundamental geographical principles and concepts.

Having laid some of this groundwork I want then to consider five case studies or vignettes: examples of work that I shall consider as part of the geography of health. I shall describe these studies, and my purpose in doing so is two-fold. First, I wish at a very early stage to introduce some pieces of research in the geography of health in order to capture the imagination. Second, I intend to use them in order to show something of the richness and diversity of geographical research on "health." There are several contrasting approaches to "doing" the geography of health, and there is no single style of enquiry that is accepted by those working within this field. I shall not say very much in this chapter about these different styles, though I hope it will be clear from what I do say that the differences exist. Instead, I shall leave until Chapter 2 the task of explaining how these five studies differ, and I shall use other examples to show how there are, broadly speaking, five alternative "modes of explanation" within the geography of health. It will emerge later that this classification is far from clear-cut, but it serves as a useful organizing device. It will also become clear from the rest of the book that by no means all approaches are widely used in studying the geography of health. Nonetheless, I think it is essential at an early stage to set out different styles and approaches.

Health and Geography: Some Fundamental Concepts

This section considers some concepts needed for a basic understanding of "health" and of "geography." They are examined separately – and in a real sense the remainder of the book represents an attempt to show the intimate connections between the two subjects. Clearly, entire books are devoted to each; my aim is simply to provide some ideas that will aid the grasp of later material.

Concepts of Health

Aggleton (1990, Chapter 1) considers several ways in which *health* may be defined. One official definition comes from the World Health Organization, who regard it as "a state of complete physical, mental and social well-being." This ideal state does not, however, assist us very much since, according to the definition, most if not all of us are unhealthy at all times! We could instead take health to mean the availability of resources, both personal and societal, that help us achieve our personal potential. Or, we might think of health as being physically and mentally "fit" and capable of functioning effectively for the good of the wider society. Linked to this is the idea of health as personal or mental "strength," fitness, or energy, or engaging in what we might think of as healthy behaviors or lifestyles (drinking alcohol in moderation, or getting regular physical exercise). Alternatively, we could think of health as a commodity, to be given or lost, bought or sold; we "invest" in health perhaps by taking out private health care insurance, and lose it when we break a leg or become ill.

Consider how you behave if you feel unwell. This might take the form of a headache or a sore throat. In the first instance you would possibly decide to manage this symptom yourself, perhaps by taking to bed or by self-medication using an over-the-counter remedy. If the symptoms persisted, or took a different form, you might consult a health professional; perhaps a nurse in a clinic, or a general physician. You do this because the symptoms represent a departure from your usual healthy state. You may be examined and tested for signs of some underlying pathology or disturbance in the body's functioning. You experience some discomfort, some pain perhaps; you feel ill. *Illness*, then, is a subjective experience. The health professional, however, is concerned to offer a diagnosis; to "identify the specific underlying pathology in the patient's body that is producing the signs and symptoms, distinguish it reliably from other possible diagnoses, and label it correctly with the name of a medically recognized *disease*" (Davey and Seale, 1996: 9). Put simply, people suffer illnesses, while doctors diagnose diseases. The doctor or physician wishes then to cure the patient of the disease; the patient will, of course, wish to be cured of any disease, but also wants to be freed from feeling ill.

Disease and illness may or may not be associated, in that it is perfectly possible to feel ill without there being any detectable biological abnormality, while the person who has been diagnosed with such an abnormality might feel quite well. For example, those who are debilitated by a feeling of complete lethargy may find that a health professional is unable to detect any obvious "cause" (and hence conditions such as myalgic encephalomyelitis, or ME – also known as chronic fatigue syndrome – may go unrecognized by doctors; it may be a condition they are not prepared to diagnose; see Aronowitz, 1998, for an extended discussion of these issues). Equally, a middle-aged man who visits his general physician for a health check-up may be feeling well but is diagnosed with high blood pressure. He arrives as a healthy person and leaves as a patient (Seale and Pattison, 1994: 16).

Since the absence of health is perhaps easier to grasp than health itself it is no surprise that we find it easier to collect data on disease and illness. Further, we find it easier in principle to "measure" disease, since we can observe and record numbers of people with a particular cancer or heart disease, for example, while illness, as subjective experience, may need recording in other ways, as we see later. We call the study of disease in populations *epidemiology*, and a substantial body of material in this book could be labeled "geographical epidemiology": the study of how disease is distributed in geographical space. Epidemiologists focus on *mortality* (death) and its causes, or on *morbidity* (sickness, which can include both disease and illness: Gray, 1993, Chapter 9). Almost always we find it sensible to calculate mortality or morbidity rates, since this allows for comparisons between populations. We also usually compute *age-*

standardized rates, thus controlling for the age structure of a population; knowing that the crude (not age-adjusted) death rate in one place is twice as high as elsewhere carries little information if we also know that there are many more older people living there. Adjustment for age structure yields so-called *Standardized Mortality Ratios*, allowing for comparisons between places that have contrasting age compositions (box 1.1).

Mortality data in the developed world come from death certificates, which also specify cause of death; this may be far from easy to establish, particularly among the elderly. Moreover, mortality is a drastic measure of ill-health! Many illnesses and diseases cause a burden to the sufferer, as well as impacts on health care systems, without leading to death. As a result, health researchers often collect morbidity data, via a number of possible routes. These can include one-

---

BOX 1.1   *Standardized Mortality Ratios*

Frequently, we wish to make comparisons of death rates between different places, but we need to allow for the fact that one area (such as a retirement resort) might have an older population than another. Since older people are more likely to die than younger people we need to ask whether the pattern of deaths in one area is independent of its age composition. We therefore "standardize" mortality rates. Such standardization is usually done by age-group where, for each areal unit, we predict or "expect" a particular number of deaths in different age categories, based on the distribution of the population by age in that area and the age-specific rates of death or disease in some wider (or "standard") population. We then calculate the ratio of observed to expected deaths, multiplying by 100. A value of 100 indicates that the death rate in an area is the same as that over a wider region (perhaps the country as a whole), a value greater than 100 suggests mortality is elevated even allowing for population structure, and a value less than 100 suggests it is reduced. Formulae are available that allow us to assess the extent to which values in excess of 100 are significantly elevated.

This method is known as *indirect* standardization and is useful in making comparisons between areas. The same principles apply to morbidity data, in which case we speak of standardized incidence ratios. We may then map standardized mortality (or morbidity) rates across the study area, though this is not without problems, as we see in Chapter 3.

Some authors use *direct* standardization, essentially the reverse of indirect standardization. The age-specific death rates for an area of interest are applied to the numbers of people in age groups of a standard population, giving the number of deaths expected in the standard population if the death rates in the area had applied. An age-standardized death rate for the area of interest may then be calculated. See McConway (1994: 90–1) for further details, including a worked example of an indirectly standardized SMR.

off patient surveys, or data from hospital consultations and related activity. We shall encounter studies based on these sources at various points during the course of this book. Such data permit the estimation of an *incidence rate*, the number of new cases occurring within a given time interval expressed as a proportion of the number of people at risk from the disease. Alternatively, we can estimate *prevalence*, the number of people with the disease or illness at any one point in time.

Without attempting here a comprehensive classification of disease, we need to draw a distinction between different broad categories. In particular, we need to distinguish between *chronic* and *acute* diseases or disease episodes. Chronic diseases are those such as heart disease, diabetes and rheumatism, which may be long-lasting and even life-long, while acute diseases are those such as myocardial infarction (heart attack), sudden stroke, or appendicitis: conditions that start abruptly, last perhaps for only a few days and then settle, though perhaps developing into chronic conditions or leading to death. Those suffering from a disease such as asthma may experience it in both a chronic and acute form, able to manage it themselves on a long-term basis but perhaps requiring hospital admission if they have a sudden acute attack. *Infectious* diseases (such as measles, influenza, and tuberculosis) are those caused by organisms that can spread directly from one person to another.

If someone is restricted in some way from general physical or mental functioning, we can speak of *impairment*. For example, chronic respiratory disease may limit one's ability to negotiate stairs, while visual impairment varies from the quite mild (short-sightedness) to the most severe (blindness). Others whose impairment confines them to a wheelchair are disadvantaged by social attitudes or poorly designed environments and buildings as well as by the cause of their impairment. Some authors (for example, Gleeson, 1999; Thomas, 1999) prefer to make a formal distinction between "impairment" and "disability," arguing that the former refers to some defective or missing body part while the latter is a socially or culturally constructed form of exclusion. What this means is that at different periods in human history, or in different geographical settings, the same physical or mental impairments might be regarded quite differently by the wider societies. For example, Gleeson (1999) suggests that in feudal societies impairment was not uncommon, but that the treatment of such individuals as "disabled" only emerged with the rise of capitalism: "Within the complex, layered dependencies which constituted feudal village life, physically impaired people were not isolated as 'social dependants' – this abject identity was a construction of the capitalist social order" (Gleeson, 1999: 97). Even in contemporary Western society many of those who are impaired may be oppressed in just the same way as other minority groups whose faces (or bodies) do not "fit."

We saw earlier how illness may trigger a visit to a general physician or other health professional engaged in "primary" care. The diagnosis may call for a referral to other, more specialized health professionals, usually in a hospital-based setting (the "secondary" sector), or even for very specialized care (perhaps complex surgery) in the "tertiary" sector. But this possible sequence of care is very much a traditional western model. *Complementary* (or alternative) medicine has grown rapidly in some western countries, however, and some health care will be delivered by practitioners such as osteopaths, acupuncturists, and homeopaths; for some of these groups, treating the person rather than the disease takes priority. Geographers are beginning to map how the use of such complementary medicine varies from place to place, and why (Curtis and Taket, 1996: 117–22; Verheij et al., 1999). In the non-western world, "*traditional*" medicine is the norm. Like complementary medicine, this emphasizes the links between mind and body in an "holistic" approach to illness and disease. Among a variety of such perspectives is Ayurvedic medicine, practiced among Hindus in India for over 2000 years. In common with other traditional forms, this emphasizes the necessary balance between three primary "humors," with therapies involving changes in lifestyle and diet, use of herbal remedies, and meditation (Aggleton, 1990: 53–9; Seale and Pattison, 1994: 18–21).

Geographical Concepts

I want to begin my brief exploration of some geographical concepts by considering *location*. I shall take "location" to mean a fixed point or geographic area on the earth's surface, somewhere that can be pinpointed by using a pair of locational coordinates. These coordinates are often latitude and longitude; for example, 51.17°N, 30.15°E refers to a location about 51 degrees north of the equator and 30 degrees east of the Greenwich meridian, while the location 23.16°N, 77.24°E is closer to the equator and further east. At a more local or regional scale it is more conventional to use a coordinate-referencing scheme specified by a national mapping agency. For example, the Ordnance Survey in Britain uses a pair of coordinates (or "grid reference") known as "eastings" and "northings." The grid reference (348211, 458826) thus identifies a unique location in Britain, and has a precision of 1 metre. Of course, few of us use such locational identifiers in our daily lives; instead, we refer to a location using an address. Such an address, when containing a postal code of some sort (known as a postcode in the UK, or a zipcode in the USA, for example), turns out to be extremely valuable to a geographer, since there may be computer-based "look-up" tables that link such postal codes to the coordinates described earlier. The uses to which this may be put are considered in Chapter 3.

What we mean by a "location" does, of course, depend on whether we are looking at health or disease in the world as a whole, or perhaps within a region. For example, if we wished to look at the way disease might be spread via air travel it would be useful to define a set of locations as the set of major cities connected via air networks. But at a regional level it would make little sense to treat Los Angeles or Sydney as a single "point" location.

Locations, then, are points or areas on the earth's surface; they do not seem to *mean* very much. However, your own home, which has an address (a location), may mean a great deal to you. A favourite theatre, or sports stadium, or woodland, or town, all identifiable locations, may well mean something to you. And if I tell you that the locations 23.16°N, 77.24°E and 51.17°N, 30.15°E, referred to above, identify Bhopal and Chernobyl, respectively, then both were etched forever on our collective consciousness as the sites of devastating industrial explosions in 1984 and 1986. Once named or labeled, these locations become *places*. Locations become places when they are charged with meaning. Until the explosions, these places were, for many of us, simply dots on a map – locations – though for those living there they had always been places. We consider the Bhopal explosion further in Chapter 7.

Places, like locations, can refer to very small areas, or be of quite vast extent. Grandfather's favourite chair, positioned to observe daily life outside the living room window, may mean a great deal to him and contribute significantly to his mental well-being; for him, it is a place. For others, particular buildings will be of enormous importance. Some stories appearing in the media, to which I referred earlier, focus on the possible closure of hospitals, and while some will object to a consequent reduction in access to health care, for others it is the symbolic quality of the institution or building that matters as much. For others still, small neighborhoods are foci of meaning, and proposals or action to alter their character, for example by locating a noxious facility in their midst will provoke opposition among those who fear it carries a health risk. More broadly, people develop attachments to places that may be cities or regions, as well as to nation states.

Places may be good or bad for health – indeed, this is a major theme running through this book. In some cases, as in the opposition to a noxious facility, the public perception of risk may be as important as any measurable impact on morbidity. As we shall see in Chapter 4, those places which are impoverished in terms of access to health-promoting resources, such as leisure and recreation facilities, will not be associated with good health to the extent that resource-rich neighborhoods are. Moreover, what happens in one place may have negative, even drastic, consequences for those living both nearby and at a considerable distance. Such "externality effects" are illustrated by the impact on health of those living downwind of a major chemical installation, or, more dramatically and over much greater distances, by the Chernobyl explosion. Conversely, other places and landscapes are considered to be beneficial to health; they are *therapeutic landscapes* (box 1.2). For ordinary, "lay" people free of any disease, however, there will be other, much more anonymous, places where we feel close to nature, where we feel secure, and with which we identify; in short, where we feel "well."

Attachment to places for some may mean separation for others. As Cornwell (whose work we examine more fully in the next chapter) notes, "where there is belonging, there is also not belonging, and where there is in-clusion, there is also ex-clusion" (Cornwell, 1984: 53). Those "attached" to a place are often at pains to object if anyone else wishes to attach themselves to it, especially if they are a different color. At the extreme, this can have severe "health" consequences for those on the receiving end of violence directed towards the "Other," those whose faces are "out of place." What this means is that places, and how we identify with them, are not simply a matter of subjective experience; "rather, such feelings and meanings are shaped in large part by the social, cultural and economic circumstances in which individuals find themselves" (Rose, 1995). This indicates that there is a danger in romanticizing the notion of a sense of

BOX 1.2   *Therapeutic landscapes*

"Therapeutic landscapes are places that have achieved lasting reputations for providing physical, mental, and spiritual healing" (Kearns and Gesler, 1998: 8). Such reputations may be built on the qualities of the physical environment, such as a source of water or a distinctive piece of topography. Or, they may rest on the qualities of buildings, such as temples. But such places are built upon reputations, and in this sense their therapeutic properties are socially or culturally constructed. People will seek out such places in order to be "cured" of a chronic disease, perhaps, or to hope for an improvement in well-being.

The concept of therapeutic landscape has been developed by Wil Gesler, whose examples include: the sanctuaries first established in classical Greece (at Epidauros, for example: see Gesler, 1993); spa towns (such as Bath in England: Gesler, 1998) based on the perceived healing qualities of local springs; or sites of deep religious significance (such as Lourdes for Catholics and other Christians: Gesler, 1996). Different people "see" these sites in different ways and at different times; for example, some claim genuine healing powers for spa waters, while for others a spa town may be simply a brief stop on a vacation. Even those who are inspired by a piece of landscape, but do not necessarily gain therapeutically from it, may notice an improvement in their mental health. Further, depending on the circumstances and our mood, even rather "ordinary" landscapes may inspire feelings of well-being.

For further details and examples see Kearns and Gesler (1998).

place. As Mohan (1998: 120) observes, "sense of place has most often and most strongly been associated with economic adversity – for instance, the instinctive collectivism of communities suffering the excesses of capitalist industrialization, such as mining settlements." Is, he asks, "the implication really that one should celebrate such conditions?" For plenty of people places are health-damaging (often in the sense of being a locus of unemployment) rather than health-promoting, and typically those with adequate resources are more likely to find themselves in the latter.

As we shall see later, geographers may choose to study health in a particular place, or they may want to make comparisons between places and study health events and outcomes in a set of places. If the latter is important they will frequently want to consider and measure the *distance* that separates places. How far are people from those facilities delivering health care? How far are people from a possible source of pollution, such as a smelter? Over what distances do diseases spread? We have already seen that locations can be pinpointed in an absolute sense; but as these examples make clear we will often need to look at where places are located in relation to other places ("relative location"). Distance, then, is something which relates one place (or location) to another. It is perhaps *the* fundamental concept in geography. How is it measured? An obvious and important measure is the physical distance separating one location from another on the earth's surface. If measuring the distance from Chernobyl to Kiruna in northern Sweden we shall need to take into account the earth's curvature to do this; however, if looking at distances between the home locations of those stricken in a city by an infectious disease we could safely ignore this curvature and measure straight-line ("Euclidean") distance. It is worth pointing out that there are other concepts of distance that may be significant in a health context. We can think of spatial separation in terms of travel time, for example, or travel cost, or people's estimates of such separation ("cognitive distance"), or the social distance that separates them (in terms of class, income or lifestyle, perhaps) from their neighbors. Measuring distances between areal units (such as health regions, counties, or catchment areas, for example) creates special problems. Sometimes we can measure straight-line distances between the centers of such zones; often we are only interested in whether one zone is adjacent, or connected, to another zone and in this case simple contiguity or adjacency serves as a measure of distance.

I have already made oblique reference to it, but I want now to introduce formally the concept of spatial *scale*. This too is quite fundamental to what follows, since while health is a property of the individual we can aggregate health events for those living in a neighborhood, city, region, or country, in order to estimate disease rates for a set of such units. We might then choose to study disease incidence in a city, or make comparisons between rates for all countries in Europe, for example. We may find that factors that explain variations in disease incidence at one scale may be quite unimportant at another. For instance, international variations may be a function of how much expenditure on health care is committed by governments, or even by differences in diagnostic or recording methods, while variations from place to place within a small region may be explicable by an environmental factor. Equally, the kinds of events that impact upon our health may operate at different scales. Local contamination of a groundwater source may have consequences for those living in a quite restricted area, while (as noted already) a catastrophic nuclear explosion may have an impact across continents. On a very different scale the quality of social relations within the home will have impacts on the health of its occupants. We shall see in what follows, therefore, that geographers span a wide range of spatial scales in studying health and disease.

Although geographers concern themselves fundamentally with spatial concepts we should not neglect to mention the importance of *time*. This is because while locations remain fixed over time (if we ignore the continental drift that takes place over geological time!) places do not. Those which are inhabited may gain or lose population, with possible health impacts. Chernobyl has been a place for as long as humans have inhabited it, though one could argue that for most of us it carried little or no significance until 1986. And time-scale matters as much as spatial scale; we may choose to study the health consequences of catastrophic, extreme events (such as the tidal waves or *tsunami* that devastate Pacific coasts from time to time, earthquakes, and other natural hazards), or we may concern ourselves more with the impacts of longer-term changes, such as global environmental change (discussed in Chapter 9). Climate affects health over different time-scales; daily hospital admissions for asthma may be elevated by climate events such as thunderstorms the day before, while seasonal change brings marked mood swings in some people ("seasonal affective disorder").

Places may be good or bad for health at different times and over different time-scales. We may, for example, be exposed to particular sources of environmental contamination at different periods of our lives, depending perhaps upon where we live and the work we do. Our "life courses" will have a major bearing on our health, and we cannot neglect the influence of our migration histories on health outcomes; indeed, Chapter 6 is devoted to this theme. Further, our health is affected by what happens to us as we move around during our daily lives, perhaps being exposed to air pollution from motor vehicles as we commute to work, to a risk of accidents in some occupations, or maybe to overt or verbal violence in domestic settings. I shall consider one way of conceptualizing the role of time in the next chapter (see pp. 42–3).

While places may change over time in observable, measurable ways, it is important to note too that our *experience* of, or *beliefs* about, them may change too. The marshlands of south-east England and other parts of Britain were considered very unhealthy 200 years ago, because of the risk of malaria (literally, "bad air;" see Dobson, 1997) while south-east England now carries a reputation as one of the healthier parts of the country. Moreover, our experience or "cognition" of time changes through our lives; for me, the last seven years have "flown by," in contrast to my teenage years; this too may affect our psychological well-being. But this experience of place also changes over very short time-scales; for example, we might feel entirely safe walking through a park during the day, yet quite threatened at night-time. The location remains the same, but the "place" changes character dramatically during 24 hours. One's fear of crime can have a very real impact on health and well-being (see below, p. 41). Further, our access to health services changes over time. We may live next door to a health center, but if it is closed for the weekend we may well have to travel much further for immediate attention, while if it closes permanently because of health service restructuring there will be longer-term impacts on access to services.

Places may also be thought of as social settings or social environments; we are literally surrounded, or "environed" by other people and features of the landscape. However, we also think of *environment* in the sense of the physical world and how it impacts upon us. Climate affects health, in both a direct and indirect sense, as we shall see in Chapter 9, while in earlier chapters we look at the impacts of environmental degradation, both of air and water, on health. Even the local geology can have health impacts. For example, goitre (an enlarged thyroid, resulting in severe swelling of the neck) in areas such as south-central Sri Lanka is thought to be due in part to low levels of iodine in water and soils (Dissanayake and Chandrajith, 1996).

The environment figures prominently in a branch of the geography of health known as *disease ecology*. Here, the argument is that one cannot understand the distribution of a disease, particularly an infectious or parasitic disease, without knowing about its relationship to local and regional ecologies – the interactions between topography, climate, water, soils, plants, and animals.

Various examples of an ecological approach figure in this book, of which malaria is a good example, since we require data on particular configurations of rainfall and temperature, as well as knowledge of animal (mosquito) and human behaviors, to predict its spatial distribution. Yet the environment has impacts on health in much more subtle ways. A case can be made for suggesting that loss of biodiversity, and the despoiling of landscapes, has a negative impact on well-being. Those who derive mental health from the enjoyment of particular landscapes may well find that others' modification of such landscapes in an environmentally insensitive way causes genuine, albeit hard-to-measure, ill-health. Modern public health sees the environment as social and psychological, not merely as physical. In this sense, then, "environment" and "place" converge to provide a spatial context for health that transcends the individual's own behavior and health outcomes.

Geographies of Health: Five Case Studies

Having set out some key ideas and concepts, I want to add some color by illustrating some work that I take to be representative of the rich variety of the geography of health.

AIDS in Uganda

Consider first the description and explanation of geographical variation in the incidence of AIDS in Uganda (Cliff and Smallman-Raynor, 1992). This study uses data accumulated up to February 1990, by which date 12,444 cases of AIDS had been reported. We must bear in mind that this was likely to have been a substantial under-estimate of the true incidence, because of under-reporting and gaps in diagnosis. The most recent figures (up to May 1997) from the World Health Organization (WHO) indicate that there have been nearly 52,000 AIDS cases reported to WHO, while WHO's own estimates suggest that the cumulative number of AIDS cases may be as high as 1.9 million (WHO Epidemiological Fact Sheet on HIV/AIDS, 1999).

The starting point for the analysis by Cliff and Smallman-Raynor was a map of AIDS incidence for the 34 districts into which Uganda was then divided for administrative purposes (figure 1.1). This shows rates per 100,000 of the population (a denominator that is itself likely to be subject to error of unknown magnitude). Broadly speaking, there are two regions of high incidence, one to the south comprising three districts bordering Lake Victoria: these are Kampala, Rakai and Masaka. Here, incidence rates are greater than 150 per 100,000, over twice the reported national rate of 74 per 100,000. A second region is to the north, where Gulu has an incidence of 255 per 100,000 and surrounding districts also have an incidence of more than 50 per 100,000. Why, then, do some districts have very high, and others relatively low, incidence?

Cliff and Smallman-Raynor consider three hypotheses. The first suggests that the AIDS virus (HIV) "trickles down" from urban to rural areas via workers

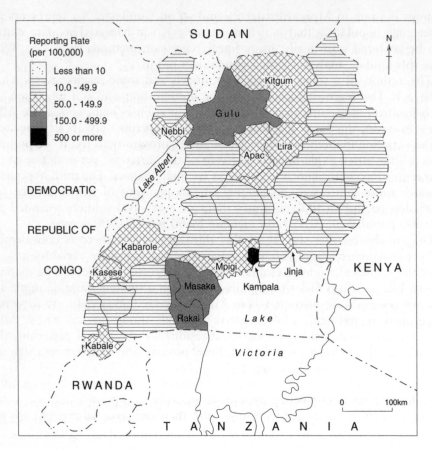

Figure 1.1   Incidence of AIDS in Ugandan districts (Source: Bailey and Gatrell, 1995)

returning home to rural areas from their urban workplaces. Areas of high incidence, it is argued, are those demanding labor as well as those supplying a workforce to urban areas; consequently, regions of high in-migration and out-migration of labor should have higher AIDS rates. This is the "migrant labor" hypothesis. A second suggestion is that HIV has spread via major transport routes and that districts which are close to one of four roads (labeled A,B,C, and D in table 1.1) are those in which AIDS rates would be relatively high; the further away districts are from one of these routes the lower the incidence. Lastly, it is suggested that the Ugandan military has played a significant role in the spread of the disease. There is certainly evidence from other African countries of high rates of HIV infection among existing and former soldiers. To investigate this hypothesis data are collected on army recruitment, drawn from different tribal groups; since these data are not collected for each of the 34 districts there is some uncertainty in deriving recruitment rates for such areas.

Although Cliff and Smallman-Raynor do not report the data, we can think of it as comprising a table or data matrix (see box 1.3). Here, for example, Y_1 represents the observed incidence rate (variable Y) in Kampala (district 1), Y_2

denotes the rate in Mpigi (district 2), and so on. Similarly, X_{21} represents the percentage population that is urban (variable X_1) in Mpigi. This data matrix can be analyzed statistically, using linear regression methods (box 1.3). What does this kind of analysis of the Uganda data tell us?

The authors fit a series of regression models and some results are shown in table 1.1. The regression coefficients for in-migration and out-migration are both positive, suggesting that where labor migration is high, so too are AIDS rates; however, this labor migration hypothesis explains only about 6 per cent of the variation in incidence ($R^2 = 5.7\%$). The "proximity to roads" hypothesis does a little better, explaining 17 per cent of the variation; yet only one of the four regression coefficients is negative, as we might expect! The third hypothesis is much more impressive, accounting for 43 per cent of the variation. The regression coefficient (0.6) indicates that, for every 100 soldiers recruited per 100,000 population, AIDS rates increase by 60 per 100,000 people.

Despite the relative success of this third hypothesis there is considerable (57 per cent) unexplained variation, suggesting that there are variables missing from the statistical model. Residuals from the regression model can be mapped (figure 1.2, p. 18) in order to explore what such a missing variable might be. Positive residuals are districts where AIDS rates are higher than are expected, given army recruitment, while negative residuals show districts whose AIDS rates are over-predicted by the model. If residuals show spatial patterning this is evidence of a missing variable; but, here, positive and negative residuals are randomly mixed across the map.

This study follows a classic model of scientific investigation in geography, considering a set of competing hypotheses in order to explain spatial variation. We begin with a mapped distribution, and then endeavor to account for the

Table 1.1 Regression results for tests of hypotheses to explain variation in AIDS in Uganda

Variable	Regression coefficient
Hypothesis 1	
Labor out-migration	23.80
Labor in-migration	42.30
$R^2 = 5.7\%$	
Hypothesis 2	
Distance from road A	4.63
Distance from road B	0.17
Distance from road C	0.08
Distance from road D	−0.33
$R^2 = 17.4\%$	
Hypothesis 3	
Army recruitment	0.60
$R^2 = 43.0\%$	

Source: Cliff and Smallman-Raynor (1992: 193)

BOX 1.3   *Regression analysis*

Regression analysis is a widely used statistical method that seeks to use one or more explanatory variables (sometimes called "independent" variables or "covariates") to account for variation in a response variable ("dependent" variable). It requires a data matrix comprising values of Y, the response, and values of one or more X variables (covariates), as follows:

District	Y (incidence)	X_1 (Urban population)	X_2 (Labour out-migration)
1 Kampala	Y_1	X_{11}	X_{12}
2 Mpigi	Y_2	X_{21}	X_{22}
. . .	. . .	. . .	. . .

It is easiest to visualize where the response variable is measured on a continuous scale, and where there is a single covariate, also measured on a continuous scale; see the diagram below. Each observation, a district in Uganda for example, is represented as a point on a graph (the set of points forming a "scatterplot"). Regression analysis seeks to pass a line through the scatterplot so that it passes as close to the points as possible. The slope of this line (the regression coefficient) shows how Y varies with X; specifically, how a unit change in X produces a unit increase in Y. This might be positive (as X increases, Y increases too) or negative (as X increases, Y decreases). The extent to which the data points cluster round the regression line is measured by a correlation coefficient ("r"), the square of which represents the proportion of the variation in Y that is "explained" by X.

The regression line has two parameters, its slope and intercept (the value of Y that is predicted when X is zero). For any observed value of X we can consider two accompanying values of Y; the observed value (Y_i), and that predicted by the regression model $(\hat{Y}_i)$. As the graph makes clear, some data points are poorly predicted by the regression line; that is, the observed and predicted values of Y are very different. The difference between the observed Y value and its predicted value is called a regression "residual" (e_i). Observations lying above the regression line are positive residuals (observed minus expected Y is greater than zero), while those below the line are negative residuals.

continued

pattern. This is done by assembling, perhaps from a variety of sources, a set of observed data, for both the "response" variable (in this case, AIDS rates) and candidate explanatory variables. If the incidence were uniform there would be no "geography;" here, clearly, there is geographical variation in incidence and this requires explanation. But, as we shall see in the next chapter, geographers aim frequently to go beyond the "static" map in order to trace the dynamics of disease spread.

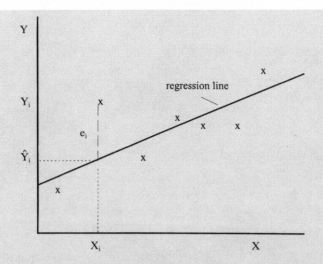

Simple linear regression can be extended readily to situations where more than one explanatory variable is available ("multiple regression analysis"). Often, more specialized forms of regression analysis are required, known as *generalized linear modeling*. For example, disease rates are formed by dividing a count of disease cases by some suitable denominator. Rather than use the rates as a dependent variable it is more appropriate to use a so-called *Poisson regression model* (McNeil, 1996).

See Robinson (1998: 93–106) for a fuller account of regression analysis.

Childhood Accidents in Huddersfield, UK

Many types of illness and death that we shall consider in this book cannot be clearly given a location of origin. This is especially true of chronic rather than acute illness and disease. It is difficult to say that a particular cancer "happened" in a particular place, or that an individual contracted multiple sclerosis at such-and-such a location. In contrast, accidents, such as those resulting from road traffic or in the home, have an obvious locational reference. We can expect, therefore, "place" to be important in an understanding of the incidence of such accidents. One study which illustrates this well is due to a collaboration between sociologists and public health specialists in the West Yorkshire town of Huddersfield (Sparks et al., 1994).

Accidents among children under 15 years of age are a major source of mortality and disability in the western world. There are strong social class gradients in rates of accidents, with children from poorer backgrounds at least five times as likely to die from accidents as those from more affluent backgrounds. Plenty of empirical data exist to support this assertion. What is needed, and what the Huddersfield team aim to provide, is an explanation of this health

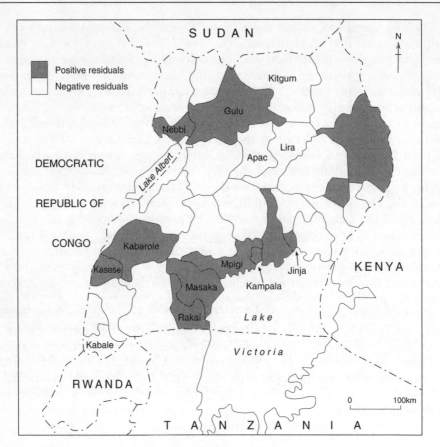

Figure 1.2 Residuals from regression model of AIDS incidence in Uganda (Source: Bailey and Gatrell, 1995)

"divide" (see Chapter 4). Explanations can then feed into government policies to reduce these inequalities; in Britain, for example, policies to reduce accidents have for some time been a key target area.

In order to explain variations in childhood accidents Sparks and colleagues examine peoples' accounts of the processes, contexts, and strategies that arise in their day-to-day health-related behavior. Specifically, they examine the perceptions and attitudes of a small number of parents in two contrasting areas: one, Holmfirth, has a low rate of accidents, while the other, Deighton, has a higher rate of childhood accidents. Holmfirth is a semi-rural area, with low (10 percent) unemployment, households that are mostly (83 percent) owner occupied and a low proportion (5 percent) of single-parent families. Deighton is a deprived urban area, with 33 percent unemployment, only 43 percent owner occupiers, and 18 percent single-parent families. In-depth interviews (see Chapter 3) were held with 18 families in Holmfirth and 14 in Deighton; half of these had had a child involved in an officially recorded accident during the previous year, while half had not.

The interviews focused on three broad topics: the parental perspectives on childhood safety and accidents, accident "events," and the development of routines and practices of safe behavior. Those living in Holmfirth see the place as generally safe, though are sensitive to traffic hazards; as one says, "I am naturally concerned but I feel my children are basically safe." They see themselves as primarily responsible for preventing accidents to their children. Those in Deighton see the place itself as fundamentally unsafe. "Bradley Boulevard is used as a race track. I often fear for my children's lives up there." Reference is made in Deighton to unfenced gardens, broken gates, unsafe roads, and risks of children being abducted; responsibility for preventing accidents is seen as a job for local government (to improve the local environment) and for the police. Few of the Holmfirth parents could recall previous accidents to their children, while all of the Deighton parents could. All parents developed rules, routines, and practises to keep their children safe; there are different "zones of control" in operation, with children being allowed more freedom to roam beyond the immediate home location as they get older. But the authors suggest that in Deighton it is more difficult for parents to "police" these zones and to feel completely secure about their children's safety.

The focus of this study is on the lay beliefs of ordinary people living in particular places. The aim is to understand a health issue from the standpoint of those directly involved, out of which may emerge more effective safety and accident prevention strategies. Opting not for a large-scale survey of numerous parents, but rather for in-depth interviews with a small sample, their view is that this method is likely to be productive in uncovering the social and psychological factors involved. These factors are mediated by the kinds of places within which people act out their daily lives.

Racial Segregation and Malaria in a British Tropical Colony

The "Protectorate" of Sierra Leone in West Africa was created in 1896, after more than a century of British control. Seventy-five years before, it had earned what was to become a long-lasting reputation as the "white-man's grave," largely because of high mortality from diseases such as malaria. (As the historian Philip Curtin [1989: 8] has shown, half the British soldiers stationed in Sierra Leone between 1819 and 1836 died.) It was the British doctor Ronald Ross who in 1898 first showed that malaria was transmitted via the *Anopheles* mosquito; for this medical breakthrough he earned a Nobel prize in 1902. Earlier, he had been appointed to the newly established Liverpool School of Hygiene and Tropical Medicine, an institution endowed by financial contributions from colonial merchants and the Colonial Office. Watts (1997: xiii) notes that: "From its very onset tropical medicine was thus an 'instrument of empire' intended to enable the white 'races' to live in, or at the very least to exploit, all areas of the globe." Ross mapped in detail (figure 1.3) the breeding pools of mosquitoes in Freetown, the capital of Sierra Leone, proposing a number of possible eradication strategies, including the draining of the pools, the use of mosquito nets, and the construction of houses for the Europeans

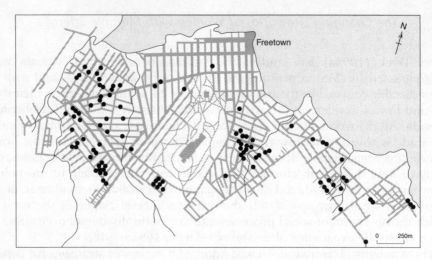

Figure 1.3 Mosquito breeding pools in Freetown, 1899 (Source: Frenkel, S. and Western, J. (1988) Pretext or prophylaxis? Racial segregation and malarial mosquitos in a British tropical colony: Sierra Leone, *Annals of the Association of American Geographers*, 78, 211–28, reproduced with kind permission of the Association of American Geographers and Blackwell Publishers Ltd)

on elevated sites (Frenkel and Western, 1988: 215). The need to protect government officers was an over-riding concern of the Colonial Office in London, and the strategy adopted was to segregate the European population from the locals by constructing a "hill station" on high ground, half a mile from the town; a distance judged sufficiently beyond the mosquitoes' reach. Since the mosquitoes were assumed to bite only at night there was no perceived problem in day-time contact between the Europeans and the locals in Freetown itself.

Frenkel and Western (1988: 224) consider this residential segregation to be a "racist" strategy, in the sense of an "almost unconscious habit of mind whereby Europeans saw themselves as unequivocally different in color and culture from fundamentally inferior subjects." Racial segregation was used as the strategy for preventing the spread of malaria among the governing population. Prevention of disease was essential in maintaining the domination of, and power over, a subject population. This spatial segregation was dictated by medical science, however, not by prejudice, at least in the published reports and memoranda issued by the Colonial Office. Nonetheless, an alternative strategy, one that upgraded all housing in Freetown, for both the African and the European populations, was either considered too expensive or, as Frenkel and Western suggest, ran counter to the prevailing colonial culture.

The general conclusion one can draw from this particular study is that the local population benefited little from the medical knowledge of how to control malaria. "Tropical medicine" (a discipline in which Ross was a key figure) emerged as a vehicle for protecting the health, and therefore productivity, of the European personnel and their families, rather than as a means of improving the health of those being colonized.

The Changing Lifeworlds of Women with Multiple Sclerosis

Isabel Dyck (1995a) has studied the "microgeographies" of women with multiple sclerosis (MS), a neurological disease of no known cause and with an unpredictable course. She focuses on the lives of 23 women, mostly in their thirties and forties, resident in the area of Greater Vancouver in British Columbia, Canada. All the women in her study were unemployed, and the primary aim of her study is to examine how they occupied and used space in both the home (hence "microgeographic") and in their local neighborhoods. As she notes, the study of how those with chronic illness cope with the problem of unemployment has been little studied. Using the qualitative research method of in-depth interviews (see below, pp. 78–82) she seeks to unravel "some of the ways in which the operation of social processes and dominant discourses over time and space shape the experience of disability" (Dyck, 1995a: 309).

The women who were interviewed adopted a variety of strategies for coping with their disability, including moving home, restructuring the physical space within the home, and reordering the pattern of domestic work. Residential relocation in order to be more accessible to amenities and services and as a means of finding a home with improved physical accessibility was essential for some. But this strategy depended upon material and personal circumstances, such as income and marital status. Those who were separated or divorced were more likely to move into more modest accommodation, and to rent rather than own their property. Within the home structural modifications to its layout could be made, but again these were constrained by tenure status and by income. Outside the home there were changing experiences of place. "Most women lived a socially and spatially constrained life, devoting much of their day-to-day activity to self-care, including rest and carefully paced household tasks" (Dyck, 1995a: 315). Again, those with little disposable income were more constrained than others. Some expressed a sense of entrapment, and worried about their safety in coping with their physical environment. Being geographically close to shops and other facilities did not guarantee access, especially where parking spaces were limited and pedestrian routes were uneven. There was diversity in how the women used their time. Some sought voluntary work and achieved a sense of self-worth in so doing; others felt bored and despondent.

Despite the different experiences and coping strategies, all the women, Dyck suggests, experienced a "biographical disruption," and all sought to renegotiate the use of space "within both the constraints and enabling resources of their lifeworlds" (Dyck, 1995a: 318). The constraints and resources exist at several levels, including (on a macro-scale) how the job and housing markets are organized and how social policy instruments are used to help some women but not others. For example, those women who had never had full-time paid employment were unable to claim disability pensions. As a result, she argues that the actions of such disabled women are "shaped within sets of constraints and opportunities which, while experienced locally, are embedded in policies and broader social and economic processes" (Dyck, 1995a: 310).

In this study, Dyck is investigating the ways in which both subjective experience and wider social and economic relations and processes operate together to shape the daily lives and actions of these chronically ill women. The women are "doubly handicapped," since they have to cope with both their disability and the (labor market) constraints that disadvantage women more than men. But the women are not passive automata, responding uniformly to the constraints of disability and economic structure; rather, they are active human agents, for whom the illness is experienced in different ways. Dyck's work is an attempt to set the experience of living with a chronic illness in the context of the wider social and economic relations that impinge on the "lifeworld". In sum, we cannot begin to understand this group of women without acknowledging both these structural constraints and the emotional and social resources that they, as individuals, bring to their daily lives. And we need to recognize that these daily lives are shaped by the experiences of, and opportunities in, particular places and localities.

The Organization of Space in Nineteenth-Century Asylums

In what might be regarded as another "microgeographical" study, the historical geographer Chris Philo (1989) has reviewed how space was organized in the Victorian asylum. His concern is to examine the rich variety of institutional plans and architectures that were designed, though not often constructed, to house those suffering from mental illness. Society's need to "close off" from public view, or at the very least to segregate, those who are outsiders of some form, different from the "mainstream," has a long history. Those who are mentally ill have, in particular, traditionally been excluded from society and placed in asylums, though as Philo observes, the confinement of "mad" people is a rather modern invention. Philo is less interested in where these asylums are located and more, in this paper, in how space within the asylum is internally structured and organized.

His starting point is the "Panopticon" or "inspection house" that the English philosopher and social reformer Jeremy Bentham proposed as a design for institutions (including asylums). The Panopticon was a building that would in principle render all inmates open to continual inspection, or at least give the illusion to those inside that they were under constant surveillance. The tightly structured internal space was supposed to promote order and peacefulness, and to remove disorder and turmoil from within the asylum. As we shall see in the following chapter, the Panopticon represents a tangible expression of institutional control, so much so that others, notably the French writer Michel Foucault, have adopted the term "panopticism" to denote all forms of social "discipline" or control. But despite the claim by some that Bentham's design was widely adopted as a model for asylums, Philo is at pains to point out that "the organization of space in many actual and proposed asylums has departed greatly from 'Panopticon' principles" (Philo, 1989: 261). His research into the historical design literature reveals a considerable variety of designs and visions regarding the nineteenth-century asylum, some of which bore no resemblance to

Benthamite principles and many of which never were translated into bricks and mortar. Of those that were, some did indeed try to afford those in charge a full view of their patients, each in a separate cell, while others sought to group patients into large dormitories and large day rooms. In Belgium, the entire village of Gheel was set aside as an "asylum without walls," divided into small units and cottages in which the "mad" were housed alongside ordinary peasant families. The village took in the mentally ill from all over the country and from overseas, and its ethos was that the integration of "lunatics" within a working family environment would speed their return to sanity. Even here, however, there was some spatial planning, with the less seriously ill segregated from the noisier and hard-to-manage patients, who were relegated to more remote houses. But this lack of surveillance and medical superintendence was anathema to those wedded to the walled and regulated asylum.

Philo (1989: 283) wants us to acknowledge that there were "all manner of different institutional and non-institutional arrangements designed to shelter, restrain, and cure the 19th-century lunatic." Some sought to bring the mentally ill under one roof and to organize space within the asylum according to a rational plan motivated by the need for constant surveillance; others adopted different models of care. As we shall see in the next chapter, this sensitivity to "difference" is a key feature of a growing corpus of work in the geography of health.

Concluding Remarks

I hope it is clear already from this first chapter that there are "geographies" of health. The vignettes have been chosen to illustrate a variety of approaches to the subject, a set of different perspectives that can be brought to bear. Some look to be more obviously geographical, in that they produce mappable patterns, whether of historical or more contemporary disease or illness. The geographical content of others may appear less obvious; nonetheless, location and place figure prominently in all. In the next chapter I want to set out in more detail what these different perspectives entail. I shall do this by laying out some of their characteristics and by describing some further case studies. In so doing, I hope to persuade the reader of the richness of approach to the subject, as well as laying some groundwork for considering particular themes in subsequent chapters.

FURTHER READING

There are a number of relevant journals that anyone interested in geographies of health could usefully consult. Of these, I would particularly draw attention to: *Health and Place*; *Social Science and Medicine*; *Journal of Epidemiology and Community Health*; and *American Journal of Public Health*. All of these have good international coverage. In addition, other epidemiological and more "mainstream" health/medical and geographical journals carry relevant papers from time to time. In Britain, the *British Medical*

Journal, the *Lancet*, *Public Health*, and *Journal of Public Health Medicine* are important outlets, as are *New England Journal of Medicine* and the *American Journal of Epidemiology* in the United States.

As far as the present chapter is concerned, an excellent discussion of some of the conceptual issues underlying health research may be found in Aggleton (1990). Also very highly recommended is the series of books on Health and Disease produced by the UK Open University; the introductory chapters in Davey and Seale (1996) and Seale and Pattison (1994) are worth reading, while the volume edited by McConway (1994) provides a superb accompaniment to both the present and the following chapter.

Jones and Moon (1987) is the classic text on the geography of health; its first chapter covers some of the introductory material dealt with here, while the second chapter looks in detail at epidemiological principles and sources of data. More recently, Curtis and Taket (1996) have identified similar "strands" of research, or approaches to the subject, and while their focus is less on the environmental, and mostly on the social, their book is an important source. Other key introductory texts and collections of essays are:

Meade, M. and Earickson, R. (2000) *Medical Geography* (Guilford Press, New York).
Learmonth, A. (1988) *Disease Ecology*, Blackwell, Oxford.
Gesler, W.M. (1991) *The Cultural Geography of Health Care* (University of Pittsburgh Press, Pittsburgh). This too considers the Sierra Leone example outlined above.
Kearns, R.A. and Gesler, W.M. eds. (1998) *Putting Health into Place: Landscape Identity and Wellbeing* (Syracuse University Press, New York).
Butler, R. and Parr, H. eds. (1999) *Mind and Body Spaces: Geographies of Illness, Impairment and Disability* (Routledge, London).

If new to geography, you could usefully start with Gould, P. (1990) *The Geographer at Work* (Routledge, London), especially Chapter 19, or with Chapter 1 in Haggett, P. (1979) *Geography: A Modern Synthesis* (Harper and Row, London). More recent edited collections, giving a good overview of contemporary human geography, are: Cloke, P., Crang, P., and Goodwin, M. eds. (1999) *Introducing Human Geographies* (Arnold, London); and Daniels, S. and Lee, R. eds. (1996) *Exploring Human Geography: A Reader* (Arnold, London).

For further, detailed, examination of concepts of distance and space see an earlier book of mine: Gatrell, A.C. (1983) *Distance and Space: A Geographical Perspective* (Oxford University Press, Oxford).

On "place" see: Jackson, P. (1989) *Maps of Meaning* (Unwin Hyman, London), or the integrated collection of essays in Massey, D. and Jess, P. (1995) *A Place in the World?* (Oxford University Press, Oxford), especially Gillian Rose's chapter. For a detailed study of the impact of the Chernobyl disaster see Gould, P. (1990) *Fire in the Rain: the Democratic Consequences of Chernobyl* (Polity Press, Cambridge).

References to each of the "vignettes" are given in the full bibliography at the end of the book. For detailed geographical accounts of research on HIV and AIDS see Smallman-Raynor, M., Cliff, A.D., and Haggett, P. (1992) *The London International Atlas of AIDS* (Blackwell, Oxford) and Gould, P. (1993) *The Slow Plague: A Geography of the AIDS Pandemic* (Blackwell, Oxford), though as we shall see there are other perspectives on the geography of HIV/AIDS. For other work on childhood accidents see Reading et al. (1999), while an introduction to geographical research on malaria is given in Learmonth (1988). Isabel Dyck has written extensively on women with disabilities; see Dyck (1995b; 1998; 1999), for further work. Last, more recent geographical work on asylums is considered in Philo (1997) and in a special issue of *Health and Place* (2000, 6:3).

CHAPTER 2

EXPLAINING GEOGRAPHIES
OF HEALTH

In the first chapter I described in some detail five case studies or "vignettes," examples of work that can be bracketed under the broad heading of the geography of health. Their subject matter differs from one to another, but the difference I wish now to highlight is that of approach to explanation. Put another way, underlying each is a different philosophical stance, and while these positions were implicit rather than overt I want now to render them explicit. It is these different philosophical approaches that lead me to speak of "geographies" of health.

What the studies had in common, however, was a "problem" deemed worthy of investigation, an issue or "riddle" (Gesler, 1991) that required some understanding or reflection. An explanation of why there was place-to-place variation in disease incidence was sought in one case; an understanding of how women coped with, and adapted to, disability in a geographical setting was required in another. None was satisfied with *description* alone – how something varies geographically; rather, each was concerned also with *explanation* of some sort. My task in this chapter is to say more about what these different kinds of explanation entail. I shall do this by setting out the key features of each, but will also introduce other examples. This will serve to fix ideas, but I believe that all of the examples considered here have something of interest to say, and will thus, I hope, serve to further draw the reader into the study of the subject, in all its variety.

I want to end these introductory remarks by saying – if it were needed – that there is no single correct philosophical stance or mode of explanation. I intend to comment on both the strengths and weaknesses of each. And while I want to discourage readers from subsequently pigeon-holing into one or other category each and every piece of geographical research in this field that they subsequently encounter, I do think it is a useful exercise to reflect upon the theoretical framework, either explicit or implicit, that seems to be employed.

Positivist Approaches to the Geography of Health

In the first chapter I used research on AIDS in Uganda by Cliff and Smallman-Raynor because it encapsulates very neatly the way in which, until quite recently, most geographers had approached the geography of health. The investigation began, in effect, with a map of incidence rates. The map is a visual representation of something that had been medically defined; its value depends on an accurate recording of all cases of the disease in question. It then becomes, in this and similar studies, a tool for asking questions. Is the spatial arrangement of rates random or not? What explains the set of above average rates that seem to cluster in parts of the study area? Additional data are sought in order to examine if the rates are to be explained by one or more variables (covariates) that may be spatially associated with the disease. Such associations are tested statistically, by setting up a series of formal models designed to test hypotheses and account for variability in incidence. In this case, the authors were rightly cautious about the quality of their data, but nonetheless it offers a basis for other researchers to confirm, or refute, the established hypothesis elsewhere.

Positivist Explanation

The study by Cliff and Smallman-Raynor is a classical piece of medical geography. It is a "scientific" study, in the sense that it adopts the methods of natural science, looking for order or spatial patterning in a set of data. The study relies on accurate measurement and recording and searches for statistical regularities and associations. It emphasizes, via mapping and spatial analysis, what is observable and measurable. Because it then seeks to establish testable hypotheses, in the same way that a natural scientist would, it has many of the characteristics of a *positivist* or *naturalistic* approach to investigation. In a health context, such approaches seek to uncover causes or "aetiological" factors, though usually the best that can be established is strong association rather than "cause." A classically positivist account would have, as its end goal, a search for laws, though weaker versions strive simply to make generalizations.

The concern in positivist approaches to the geography of health is usually to detect areal pattern or to model the way in which disease incidence varies spatially; people with the disease only appear as numbers that compose spatially varying disease rates. Location and spatial arrangement matter – indeed, these are the crucial variables in *medical* geography – but "place" is incidental. Certainly, the individuals do not speak to us; what a disease such as AIDS means to them, how they cope with it in daily life, and what they think might have caused the disease is not a matter for "scientific" investigation. Let me emphasize that this does not mean that a positivist does not collect individual-level data. Far from it. Individual-level data may well be collected, on both the disease and other characteristics of such individuals (perhaps their age, sex, residential history, income, smoking behavior, and so on), and used to sort out what factors discriminate those with the disease from those without. This may be done either via survey work – what we can refer to as "social positivism" – or in the laboratory, where "bio-medical positivists look for causes in micro-organisms and in anatomical and physiological abnormalities" (Aggleton, 1990: 75). But a positivist medical geography typically involves mapping disease data and then striving to explain the spatial distribution.

In health research in general, a positivist approach also involves frequently adopting a biomedical perspective. Here, the body is seen as a machine that may not be in good working order and needs "mending." What matters is to investigate – and, for the geographer, to investigate in a spatial setting – specific diseases that have one or more specific causes. The individual is a rather anonymous person whose features and characteristics can be ticked off on a check-list. Critics of this perspective argue that it is reductionist (the individual is "reduced" to a collection of body parts and behaviors); even if "lifestyle" factors (such as diet, stress, and social support) or the effects of air and water pollution are recognized, these are attributes to be attached to individuals and used in a statistical analysis. Positivist approaches, in general, rely on the use of quantitative, usually statistical, methods, often sampling from a wider population and seeking to generalize from the sample to the population. A corollary of this is, where possible, to have the sample as large as possible, since this strengthens the conclusions.

That location and distance are key variables in a positivist geographical approach is seen not only in the study of health and disease but also in the study of health care and its delivery. Rational planning demands the organization of services on an efficient basis, such that the costs of overcoming distance are, in the aggregate, minimized. Here, too, the search for order is paramount, a feature which some writers have traced back to the philosophy of the Enlightenment in the late-eighteenth and early-nineteenth centuries, where the Enlightenment or modernist "project" was to impose order and to eliminate disorder.

Further Examples of Positivist Approaches

To illustrate further the broad range of essentially positivist accounts I shall draw on a number of other studies. First, I shall examine further work on AIDS, research conducted in the developed rather than the developing, world. I then consider some of the large body of work on disease spread or diffusion. All are studies drawn from a potentially large set that adopt essentially positivist approaches. I shall then examine one study that is concerned with health care planning in a geographical context.

Gould and Wallace (1994) have researched the spread of the human immunodeficiency virus (HIV) through mapping the cumulative incidence of AIDS cases across the US and also analyzing the number of cases reported in the New York region. At the national scale, data for 1982 show concentration in the major urban agglomerations, followed in later years by spread to smaller cities via a process known as *hierarchical diffusion* in which the patterns appearing in subsequent maps are structured by the flows of people, along the major transport arteries, initially among the largest urban centres and then to smaller towns. Later, more local processes of *contagious diffusion* appear to operate, in which physical proximity matters more than the leapfrogging of geographical space and spread down the urban hierarchy. This contagious diffusion is "driven in large part by the extensive daily commuting fields" (Gould and Wallace, 1994: 107). The argument is that since the main form of HIV transmission is via sexual

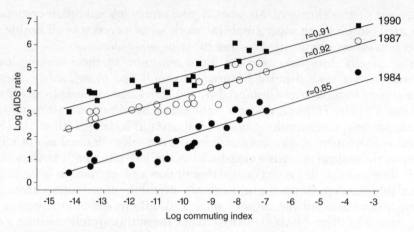

Figure 2.1 Relationship between AIDS rates and commuting in New York City region (Source: Gould, P. and Wallace, R. (1994) Spatial structures and scientific paradoxes in the AIDS pandemic, *Geografiska Annaler*, 76B, 105–16), reproduced with kind permission of the Swedish Society for Anthropology and Geography

or needle contact, these personal contacts (for which data are rare or impossible to obtain) "reflect the massive flows of interaction at much greater geographic scales" (Gould and Wallace, 1994: 107), although later they speak of the spread as being "forced" by flows of commuters. The importance of commuting flows is confirmed by plots of AIDS rates against an index of commuting for the 24 boroughs and counties of the New York Metropolitan Region (figure 2.1).

   Gould and Wallace anticipate one criticism of their research – the neglect of the human actor or voice – when they acknowledge that some researchers might prefer more autobiographical accounts of disease spread and that others might feel that poetry and other mediums of expression can be more illuminating. "But in the same way that the scientist, *qua* scientist, can say little about the individual, so the poet can write little that will help us formulate public policies. . . . Sonnets, no matter how empathetically wrenching, do not stop viruses, and do not help us to look ahead" (Gould and Wallace, 1994: 105). "[B]ecause we pose questions ultimately of cause and effect, or in other words questions of mechanism, 'social physics' is inevitably at the heart of any scientific investigation" (Gould and Wallace, 1994: 105). In their avowedly naturalistic approach there is no need for those affected by the disease to speak out. Nor is there much space in their account for consideration of the broad structural societal forces that may also cause disease spread, though they do refer to the social marginalization of those intravenous drug users living in unrelieved poverty. The study is a classic example of a positivist geography of health, one that has established a convincing association between disease incidence and an aggregate social variable. Yet it seems to me that although, in this large metropolitan region, flows of commuters are very highly correlated with AIDS rates, the causal mechanisms are missing. While the authors refer to a newspaper article which suggests that

70 percent of the clients of Manhattan prostitutes are suburban commuters, there needs to be much more empirical work – not necessarily all positivist in conception – that illuminates this and other possible causes.

Other geographers have worked more on some of these possible causal mechanisms, via both detailed historical research and more formal modeling of the diffusion process, as instanced by the series of monographs and papers produced by Peter Haggett, Andrew Cliff and their colleagues over many years. As one example, consider the spread of influenza in Iceland (Cliff et al., 1986). Iceland is chosen by the authors as it is a quite self-contained island setting, with superb medical records going back to the late-nineteenth century, all of which allow for detailed geographical description and analysis. At least 30 well-defined influenza epidemics, some involving several thousand cases, others a few hundred, have affected the country since 1900, the population having grown from about 78,000 to 210,000 in 1970 (when the authors terminate their study). In early decades the role of shipping is highlighted, with the 1918 epidemic triggered in particular by one trawler arriving in Reykjavik from Copenhagen. Communication bans and the quarantining of "infectives" were soon imposed, with the result that the epidemic was contained successfully in south-west Iceland. In the 1937 epidemic, spread was via both hierarchical and contagious processes. From Reykjavik it spread down the urban hierarchy, via local shipping movements and, once established, moved outwards from towns into adjacent rural areas, facilitated by social and business gatherings (such as a farmers' conference). In 1957–8 a new strain of the virus ("Asian influenza") arrived from overseas via air as well as sea routes, this time reaching the eastern coast. For the authors, a mixed hierarchical-contagious diffusion model seems generally applicable to all the major epidemics they consider.

Cliff et al. (1986: 168–205) identify a number of "statistical regularities" in both the temporal and spatial spread. One, for example, concerns why influenza appears in Iceland as a series of regular epidemics rather than being endemic (occuring continuously). The explanation lies in the sparsity of population. If we plot the percentage (logarithmically transformed) of months between 1945 and 1970 in which medical districts report influenza, against population size (also logarithmically transformed), we see a quite strong linear relationship (figure 2.2). This is described by the following regression model:

$$\log T_i = 0.85 + 0.38 \log P_i$$

where, for the i'th district, T_i is the percentage of months with reported cases and P_i is population size. If we give T the value of 100 per cent (that is, a district has cases every month) and solve the equation for P, we obtain a value of about 110,000. This is then the threshold for endemicity. Even the capital of Iceland, Reykjavik, has too small a population to sustain the disease continuously.

But Cliff et al. (1986: 207–59) go beyond these statistical descriptions to build mathematical models of the diffusion process, with the longer-term view of forecasting the future spread of the disease. Several models are explored, the simplest of which are based on an accounting framework, expressing the number

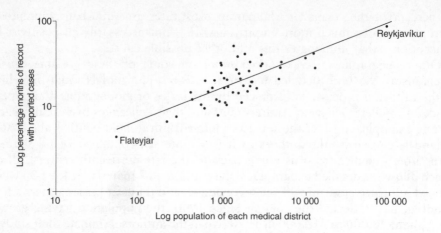

Figure 2.2 Relationship between number of months with reported influenza and population size in Iceland, 1945–70 (Source: Cliff, A.D., Haggett, P., and Ord, J.K. (1986) *Spatial Aspects of Influenza Epidemics*, p. 169, reproduced with kind permission of Pion Ltd)

of people infected at any point in time as a function of: the number of infectives in the previous time period; newly emerging cases; those removed by being no longer infective; those "arriving" (via birth or immigration); and those "departing" (via death or emigration). Models for more than one district assume that the probability of contact between an infective in one district and a susceptible in another depends upon the distance separating them. However, the models are "data-hungry," ideally requiring daily data since the disease has a serial interval (the time between the onset of symptoms in one person and in another person directly infected by the first) of only 4–5 days; typically, only monthly data are available.

These studies of spatial diffusion form, along with others, a substantial body of knowledge relating to the spatial and temporal spread of infectious disease. They involve careful recording of cases, statistical description of the data, and subsequently mathematical and statistical modeling of the process. Attempts may then be made to forecast disease spread, with the longer-term goal of devising strategies to control this spread. But those infected are anonymous individuals, aggregated into areal recording units, and it is left to other approaches to let those who are infected by, or affected by, disease spread have a voice.

Before considering alternative approaches, note that there are positivist approaches to examining the provision and delivery of health services. Given limited resources, where should we locate health services? How should patients be allocated to particular health centres or hospitals? The set of tools used here are known as *location-allocation* methods (see Thomas, 1992 and Bailey and Gatrell, 1995, Chapter 9 for an introduction). In a developed world context, Smallman-Raynor and his colleagues (1998) have demonstrated the usefulness of location-allocation modeling in planning the location of cancer units in a part of central England. In the developing world Hodgson (1988) has emphasized

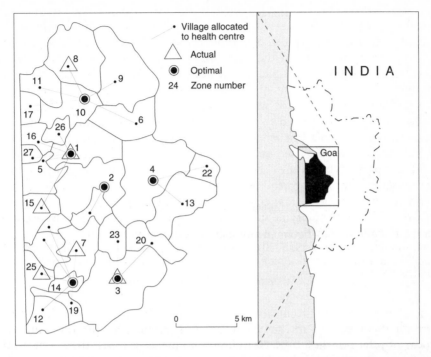

Figure 2.3 Optimal location of health centers in West Goa (Source: Bailey and Gatrell, 1995)

the hierarchical nature of the planning problem; in other words, the need to locate both basic health services, but also centers offering a wider range of health and medical care. Particular location-allocation models propose different location criteria, typically minimizing the aggregate distance traveled from place of residence to the health center. The sizes of the centers are important, however, in that they may be taken to represent, albeit in a rather crude way, their "attractiveness;" other things being equal, people will travel further to a larger center. Models are therefore needed that maximize net benefit to the patient, locating a set of health centers in such a way as to ensure that they are as close as possible to such patients. Hodgson applies a hierarchical model to the location of health centers in a region of Goa, India (figure 2.3).

Social Interactionist Approaches to the Geography of Health

We have seen how positivist accounts focus on the observable, the measurable, and the generalizable. Individual characteristics might be recorded, but nowhere is there a concern for individual meaning. In a health care planning context, the attractiveness of health centers is represented by easily available data on bed spaces, or number of health workers, for example, rather than by the more intangible features that are likely to be important to people. Consequently, other researchers wish to emphasize what is less readily measured and quantified; the subjective experience of health and illness. I shall, following Aggleton (1990), refer to these approaches to the study of health and illness as *social interactionist* (some use the term *social constructionist*). They are so called, because meanings are *constructed* out of the interactions (which may be conversations and encounters) that we have with each other in day-to-day life. Geographers have for many years referred to these perspectives as "humanistic," since they address implicitly human beliefs, values, meanings, and intentions. In a health context, such approaches see people not simply as collections of possibly diseased body parts. Nor do they see people as merely passive recipients of knowledge about health and health care; rather, people are continually engaged in the construction of such knowledge (Stainton Rogers, 1991). A corollary of this is that the views of ordinary people (referred to as "lay" or commonsense views) have as much status as those of the health professional.

Social Interactionist Explanation

In social interactionist accounts, therefore, the emphasis is on the meaning of the illness or disease to the individual and the task for the researcher is to uncover or *interpret* these understandings and meanings that make it "rational" to act in a particular way; in other words, to see things from *their* point of view. For example, not going to have a child immunized may be irrational to the health professional yet perfectly logical to the parent, while the woman who decides not to attend for breast screening may too be behaving entirely rationally in her own terms.

By extension, social interactionist approaches study small numbers of people; geographical versions will typically study small communities or neighborhoods, rather than studying large areas. This is because the experience of place is more important than the accurate recording of large numbers of locations, or the pigeon-holing of people into a fixed set of areal units. The methods used are essentially qualitative rather than quantitative, and the ultimate goal is empathetic understanding and explanation rooted in the social, rather than the natural, world.

Such approaches are, as one might expect, subject to criticism from positivists, who argue that the verifiability of results is hard to obtain, and any conclusions drawn from small numbers are difficult to credit; for a positivist, the objective matters more than the subjective. For others, the emphasis on "human agency" means that, like positivist accounts, the social interactionist neglects wider structural influences on health.

Members of the Huddersfield team who conducted research on childhood accidents are quite explicit in wishing to move away from "the dominance of bio-medical based large-scale research" (Sparks et al., 1994: 440) in which aggregate data are used to explore statistical associations between measures of illness and various social and demographic variables. For them, a limitation of this kind of work is that it tends to merely describe rather than explain. They want a greater *understanding* of the social processes that are involved in shaping health-related behaviors and outcomes.

Further Examples of Social Interactionist Approaches

Another example of research that can be bracketed under this heading is that on the experiences of local residents living near waste disposal sites; specifically, landfill sites and an incinerator in southern Ontario (Eyles et al., 1993). This work parallels a more classical epidemiological survey, uncovering the psychosocial effects via a series of in-depth interviews and focus groups (see Chapter 3 for further discussion of these techniques). Individuals were selected from among those who had, and had not, expressed concern about the waste disposal facilities, giving a total of 39 interviewees over the three sites. Only one interviewer was used (thereby eliminating one source of possible bias), and the interviews (lasting between 20 and 90 minutes) were taped and transcribed. A checklist of topics to be covered included: attitudes to home, environmental contamination and quality of life, and more specific feelings about the local facility. The results depend on whether the site is a landfill or incinerator. Those living near the latter complained about air pollution, with some worrying about short-term health impacts and others about longer-term damage, though since the incinerator is located in a highly industrialized area it is difficult to separate out the effects of the facility from those of other sources of environmental contamination. Those living near one landfill site expressed few concerns about its impact on water quality or human health; for those close by the problems were of debris and noise, while for those further away it was the additional traffic generated that caused concern. At the second (at the time, planned) landfill site there were more concerns about possible contamination of water supplies. But, to the authors' surprise, associations between environment and health were not so very prominent. Rather, anxieties were latent and concerns were more about the lack of control over siting decisions than about observable health impacts.

The authors justify their qualitative study as being complementary to the epidemiological companion paper (Elliott et al., 1993), arguing that it enriches the survey results (Eyles et al., 1993: 810). A positivist would question whether a very small number of interviewees is representative of the local populations, and worry about whether their concerns can be pinned down as due exclusively to the waste sites, rather than to some other possible sources of environmental pollution. This is countered by the argument that qualitatively based research such as this does not set out to make generalizations. We shall see something of the competing claims of quantitative and qualitative research in the next chapter.

Few doctoral dissertations on the geography of health can have had as much impact as that of Jocelyn Cornwell, whose resulting book (1984) is a detailed case study of ordinary people's accounts of health and illness, and their experience of health services, in Bethnal Green, part of East London. Cornwell's approach is anthropological and ethnographic, and she does not see "health" as a field around which one can put sharp boundaries. To that end, she is as much concerned with understanding aspects of daily lives that are, on the surface, less directly health-related: housing, work, and family life, for example. Her method is to conduct in-depth interviews with 24 people, drawn through a "snowball" process whereby the first contact spawned subsequent names to be recruited to the study. The interviews, based around a core schedule of topics and conducted over numerous sessions with the same people, provide a level of detail that no survey could ever hope to match. Cornwell conducted multiple interviews with her informants, since a single one might reveal only "public" accounts (in a sense, what the respondent thinks the interviewer might wish to hear), whereas a succession of interviews begins to tease out the "private" accounts, the deeper thoughts and feelings emerging from the respondents. The interviews were, as far as possible, led by respondents' own concerns, since one of the dangers of survey research (and some interviews) is that the relationship between interviewer and interviewee is very hierarchical, with the "expert" merely gathering as quickly as possible the data needed to advance the research. As Cornwell says, the sample is not statistically representative, but is "typical" in the everyday sense that the people's lives "faithfully reflect the history of social and economic life in East London over the past eighty or more years" (Cornwell, 1984: 1). As with Eyles' work, this marks a very clear separation from a positivist approach.

The distinction between public and private accounts is important, and often conflicting. Public accounts emerge from answering questions; private accounts emerge from the telling of stories about personal experiences. For example, public accounts stressed the almost romantic sense of place – the strong sense of community and high quality of social relationships. By contrast, private accounts emphasized the negative aspects – the fights, the arguments, and lack of concern for others. Similarly, public accounts of family life stressed the "unity" of family life, the virtues of hard work and the devotion of women to family welfare; private accounts uncovered the rifts, the disharmony, and the stresses of living together. This emphasis on family and working life is important, since it acts as the context for accounts of health and illness.

As far as public accounts of health and illness were concerned, the respondents wanted to assure Cornwell that illnesses were "real." They also made light of their own ill-health, being reluctant to be labeled as someone with poor health. Good health is morally worthy, while illness is not, unless it bears the stamp of official approval by diagnosis, in which case it is beyond the power of the individual to control it. Private accounts implicated poor working environments, such as the problems of deafness and discomfort caused by driving older lorries (trucks) and the back problems caused by handling heavy goods. Public theories of disease causation tended to be the interpretations of "medical" opinion by ordinary people; they frequently suggested that disease and illness were beyond individual control and that the person affected was not to blame. Private theories were more complex, suggesting multiple causes of disease and illness drawn from events throughout life.

As a final example of this style of research Kearns (1991) adopts what he refers to as a "humanistic" understanding of place, in order to examine the way in which health care in a quite isolated part of New Zealand (the Hokianga Special Medical Area on the North Island) contributes to a sense of place. He argues that this can occur in two ways. First, that particular forms of service provision can enhance the community's sense of well-being, simply by being in possession of good facilities; and, second, by enhancing the level and quality of social interaction in that community. His study area, Hokianga, has a scattered and quite deprived population, and a government initiative has sought to attract primary health care workers into the area and to provide free care. There are nine clinics in the area, and one hospital. Kearns interviewed health professionals in these sites, and acted as a "partial" participant observer (see Chapter 3), being "attentive to the ensuing conversations as patients and others gathered" (Kearns, 1991: 525). The topics of conversation were later categorized. Community concerns (such as the restructuring of public services) were most common, followed by health-related issues and family well-being. The clinics are therefore used as public places to gather, interact with others, and swap information; their role as places are at least as important as their role as treatment centers. Kearns contrasts these settings with those in urban areas that are likely to be more "socially sterile." For Kearns, the spatial arrangement of facilities scratches the surface of a (social) geography of health; it is what goes on in such places that can add to – or perhaps detract from – the social fabric and a sense of place.

Structuralist Approaches to the Geography of Health

The work by Frenkel and Western that we considered in Chapter 1 sought to identify some of the political and historical factors that shaped disease and health in nineteenth-century Freetown in Sierra Leone. As we show in this section, related studies suggest that the underlying causes of disease are embedded in political and economic systems. Explanations are not to be sought at the individual level – for example, the kinds of "unhealthy" behaviors they adopt. Instead, it is the broader social context that matters. As Turshen (1984: 11), has it, sickness lies not in the body but the body politic. Because of the stress on these macro-scale social, political, and economic structures, this style of approach is often referred to as *structuralist*, or alternatively as a *political economy* perspective.

Structuralist Explanation

Structuralist approaches derive much of their impetus from Marxist theories of oppression, domination, and class conflict, where inequalities are embedded in society. Such theories take a variety of forms but, put simply, they propose that economic relations and structures underpin all areas of human activity, including health and access to health care, and further that the economic "determines" the social – there is no scope in a rigidly Marxist account for human intention and free will. It is clear from this, therefore, that such a perspective is antithetical to that of social interactionism. Human agency is absent from classically structuralist accounts.

Some authors offer an explicitly Marxist interpretation of health and health care, especially in the context of late-nineteenth- and early-twentieth-century colonialism (Ferguson, 1979). As we shall see below, the emphasis in these contexts was very much on curative medicine rather than on preventive (public) health. There are two reasons for this, according to Ferguson. One is that the expertise developed by the doctor over several years' training produces a commodity (medical skill) that has an exchange value which is realized by treating those able to afford the private care on offer. The second point he makes is that since poverty is the main cause of ill-health, and poverty results from capitalism, there is little incentive among those controling, and working in, the health care system to attack these root causes, since more money is to be made from providing medical cures than reducing poverty and preventing disease and ill-health in the first place. Whether in the developing or the developed world, biomedicine emphasizes cure rather than prevention and is keener to promote "high technology" approaches to health care, even in quite rural developing countries where this is wholly inappropriate. Medicine thus serves to perpetuate social inequalities and to widen the gap between the rich and poor; it does nothing, according to the political economist, to reduce these disparities. Health improvements are contingent on overcoming such dependency and on transforming the economic system from capitalism to socialism. Unfortunately, there is scant evidence that those living in many former socialist states had very different health outcomes, or that inequality was less than in capitalist societies.

There are, however, other deep structures embedded in society that are based on conflict and power relations, of which the most obvious is the role played by male power (patriarchy) in structuring women's health. A broader reading of structuralist explanations would therefore want to see a wide-ranging emphasis on conflict, or power relations, whether this be between social or ethnic groups, between men and women, between those owning the means of production and those employed as laboring classes, or between societies. Some writers (Gerhardt, 1989; Stainton Rogers, 1991) refer to *conflict* or *dominance* theories, embracing all forms of economic, imperialist, and patriarchal domination. As a result, the studies examined in this section look at links between imperialism and both the spread of disease and the nature of health care delivered under colonial rule, before turning briefly to work that is shaped by gender-based conflict.

Further Examples of Structuralist or Conflict-based Approaches

We saw earlier how one kind of approach to the spread or diffusion of disease was based on an essentially positivist approach. However, to explain the impact of diseases such as smallpox on native South and Central American populations in the sixteenth century we need to look at the practices of Spanish and British colonialists.

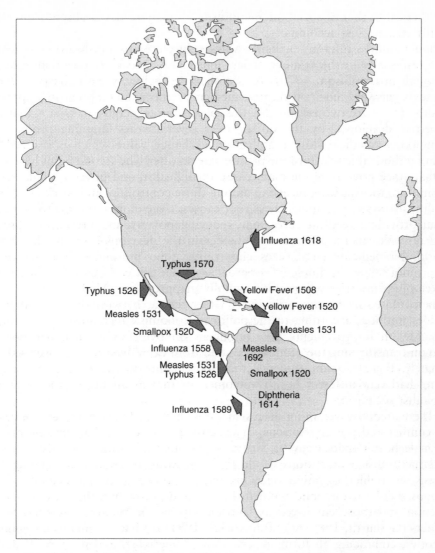

Figure 2.4 Sources of epidemics in Americas during the sixteenth and seventeenth centuries (Source: Brothwell, D. (1993) On biological exchanges between the Two Worlds, in Bray, W. (ed.) *The Meeting of the Two Worlds: Europe and the Americas 1492–1650*, Oxford University Press, Oxford, with slight amendments), reproduced with kind permission from the *Proceedings of the British Academy* (1993, Vol. 81)

Many epidemic diseases (such as smallpox, measles, yellow fever, typhus, and influenza) can be associated with the Spanish conquests of Peru and Mexico in the early-sixteenth century, and some authors (Brothwell, 1993) consider that they did at least as much as the Spanish military to dismantle the Aztec and Inca empires (figure 2.4). Such populations had no natural immunity to the diseases. One set of population estimates for central America suggests that while there were 25.2 million living in 1519, the slaughters wreaked by conquistadors in 1521 and, later, the 1531–2 and 1545–7 smallpox epidemics, had reduced this to 6.3 million by mid-century. The population had dropped to just one million by the early 1600s. After the collapse of the Aztec state following Cortes' 1521 campaign, smallpox diffused along trading networks down into South America, devastating the Inca population in 1524–5. One hundred years later, the Andean population had dropped to just 7 percent of its level in 1524 (Watts, 1997: 91). Smallpox and other viruses were spread widely by those recruiting slave labor to work in silver and other mines, where appalling living conditions encouraged disease spread.

Later, and much further north, the Pilgrims arriving in 1620 were able to observe that "the good hand of God favoured our beginnings . . . in sweeping away the great multitudes of the Natives by the Small Pox" (in Watts, 1997: 93), while in 1763, in order to wipe out further resistance among the Ojibway native Americans around Lake Superior, the British Army commander (General Amherst) oversaw the distribution among them of blankets infected with small-pox. The settlers and military in north America, like the Spanish conquistadors, were less affected, having acquired some exposure, and therefore lifelong immunity, to milder endemic forms of the virus back home.

It would be naive to imagine that native populations in the Americas lived in idyllic conditions, free from any illness. We simply have little or no record of what epidemics swept these populations before the colonialists arrived (Arnold, 1988). But what is certain is that the infrastructure imposed by colonialism, new trade routes, and communication networks aided the spread of micro-organisms and disease vectors that transmitted other diseases. Until the emergence of this infrastructure, poor communications and the "friction" of distance served to quarantine the diseases. And since rates of spread of infectious disease are greater in more densely populated areas, the diffusion of diseases such as smallpox among native Americans was aided not only by trade and labor recruitment, but also by forced resettlement into larger population groupings – whether mining compounds, plantations, or missions – for the purposes of more effective administration and use and control of the labor force. Arnold (1988) speaks of colonialism in this context as "biological warfare" or, at the very least, a major health hazard for the indigenous peoples of native America.

This is simply one brief illustration of the impact on health brought about by colonialism. Others merit examination. "The manner in which plague travelled from Hong Kong in 1894 to Bombay in 1896, to Cape Town in 1900 and Nairobi in 1902, and then to West Africa a decade or so later is indicative of the new facility of disease transmission opened up by modern trade, transport and imperial ties" (Arnold, 1988: 5). The deep-rooted structural determinant was the imperative to extract resources from a periphery in order to promote growth and prosperity in the center, but improvements in transport facilitated the diffusion of disease.

In a detailed study of the *political ecology* of disease and ill-health in a single East African country, Tanzania (formerly Tanganyika), Turshen (1984) too shows how the health of the indigenous population deteriorated in the presence of colonial rule and how the economic transformations brought about by colonialism and capitalism impacted on disease. For Turshen, as for others (Ferguson, 1979) the causes of disease are rooted in political and economic systems rather than individual lifestyles and behaviors. However, I want to dwell less on these aspects – some of which we have already examined – and more on the health care system that emerged under colonial rule (initially, under the Germans between 1884 and 1918, then under Britain until 1961) and, following independence in 1961, under the new socialist government led by Julius Nyerere. This provides a useful contrast with more positivist geographical analyses of health care systems.

Before the "scramble for Africa" in the later-nineteenth century, Christian missions had been established in East Africa, many of which introduced western medicine. This emphasized the cure of disease rather than preventive medicine, and there developed a clear association between medicine and conversion. Archdeacon Walker wrote in 1897 that: "I regard the medical work from its missionary aspect. . . . I consider how far it is likely to aid our work, not how much suffering will be relieved" (quoted in Ferguson, 1979: 319). After the partition of Africa Germany took control of what was known as German East Africa (later, Tanganyika) and established army hospitals to serve local garrisons. Under both German and early British control separate facilities were created (either separate hospitals or separate wards) for Africans, Asians, and Europeans, again with the emphasis on curative and surgical medicine. The contrasts are apparent from a description by the Principal Medical Officer in 1920. The European Hospital in Dar es Salaam held 50 beds and had "a separate maternity section, well-fitted X-ray room . . . and room for the examination of eye cases, spacious operating theatre, outpatient department and quarters for the nursing staff and Medical Officer." The Sewa Haji Hospital, built for Africans and Asians, and also with 50 beds, was "a curious rambling collection of buildings of which the administrative block is the outstanding feature" (in Ferguson, 1979: 326). The ratio of beds to population served was 1 to 10 in the former and 1 to 400–500 in the latter.

These health services were superimposed on the traditional health care system, essentially a group of practitioners whom the earlier missionaries had sought to discredit, since their traditions ran "counter to the belief that Victorian civilization was the acme of human achievement" (Turshen, 1984: 145). Turshen sees the active participation of local communities in the healing process and prevention of illness as a possible political threat to colonial autonomy and the social control that western medicine provided; this was another reason for discouraging traditional practices. Until independence in 1961 the emphasis remained on cure rather than prevention, especially among the urban labor force. Among women and children living in rural areas little attention was given to meeting nutritional requirements, though this was not the case among laborers; an official government report in 1936 confirmed that "undernourished natives cannot be expected to, and are not capable of, working really efficiently" (in Turshen, 1984: 151). The emphasis given to an urban-based curative system thus produced a very uneven distribution of health care.

After independence, and after the firm commitment to socialist development in 1967, the health care system changed dramatically, as table 2.1 suggests.

Table 2.1 Health care provision in Tanzania

	1960	1977
Physicians	425	727
Tanzanian physicians	12	400
Hospitals	99	141
Hospital beds	11 160	19 970
Rural health centres	22	161
Dispensaries	990	2 088

(*Source*: Turshen, 1984: 193)

From 1972, the government began to redirect resources towards rural health care delivery, recruiting part-time village medical helpers. As the table confirms, primary care, "lower level" facilities grew more than the secondary sector. However, as OXFAM's medical advisor in Tanzania noted in 1982, lack of resources means that "dispensary staff provide inadequate treatment, are very rarely supervised, very rarely visit the four or so villages they should cover, and are often demoralized" (quoted in Melrose, 1982: 21). Turshen regarded health care in 1970s Tanzania as still essentially "neocolonial," a system that is dependent on the international community – whether on foreign aid for health, on the employment of foreign health care workers, or on the control of drug distribution by multinational pharmaceutical companies.

There are, of course, dangers in "blaming" all health problems on colonial influence and in inferring that independence brought about change. In Tanzania, for example, Nyerere imposed the collectivization of population into villages ("*ujamaa*") well into the late 1970s and the shift away from curative medicine only began ten years after independence. We should not romanticize too much the socialist model he imposed, since the movements of population (up to 10 million persons) to the *ujamaa* were often coerced, democratic rights were curtailed, and elected officials were often replaced by government officials (Coulson, 1982). More recently, the country has become less explicitly socialist in orientation and more pluralist and liberal. Whether the continuing health problems in the country are due more to lingering effects of colonial influence, the continuing impacts of "neocolonialism," or poor management by those running health services in the country, is an argument that will continue to be debated.

Another "conflict-based" approach is a *feminist* perspective which argues that historical relations between men and women are basic to an understanding of society; power is divided unequally, with men the dominant group. There are still relatively few examples of this style of approach in the geography of health, although health research in general has made much use of a feminist perspective (for instance, in looking at women's experiences of antenatal care). But such research cannot possibly be "forced" into a simple box; it draws as much on social interactionist as on the other approaches to scholarship presented here. However, since it fundamentally opposes a masculinist approach to knowledge I consider an example of such work here.

This work relates less to an obvious health "outcome" (mortality and morbidity) and more to a problem in the broader sphere of public health, namely the fear of personal attack. There are good grounds for considering fear as an entirely legitimate component of well-being, or lack thereof (Pain, 1997). What do we know about its geography? Among other writers, Madge (1997) and Pawson and Banks (1993) have considered this issue, from a British and New Zealand perspective respectively. Madge conducted a survey of the use of public parks in Leicester, finding that three-quarters of the women surveyed said that fear of mugging or sexual attack restricted their use of parks. Such fear was most common among African-Caribbean women, younger women, and those who were unemployed. Such women avoided large open spaces, unlit areas, and areas with extensive undergrowth and trees. Madge points to the increasing neglect of, and under-investment in, public parks in Britain as being partly to blame. Spending on urban open spaces in Britain fell substantially between 1981 and 1991, with a consequent reduction in repairs to lighting, and in supervision by park wardens whose presence might deter threatening behavior. Clearly, structural factors including both potential threats from men and under-investment in the public sphere shape the way in which women are excluded from some public spaces.

Pawson and Banks (1993) also look at the geography of fear, and at the distribution of reported rapes, in Christchurch, New Zealand. Many rapes are reported as taking place in the victim's, or attacker's, home rather than in public places, but the latter are nonetheless a significant set of locations for such attacks. Nearly three-quarters of women in low-status neighborhoods feel unsafe in their own homes, and over half of them are reluctant to venture into the local neighborhood alone at night. Those living in such areas are less likely to be able to afford the home security measures that would provide some guarantee of safety. Both studies indicate that the fear of, or actual, rape and other sexual violence represents one manifestation of patriarchy; this fear gets translated into constraints on spatial behavior and as such represents a major source of structural inequality within society. I comment further on such inequality in Chapter 4.

Structurationist Approaches to the Geography of Health

We have seen how one set of alternatives to positivist explanation in the geography of health is to give much greater weight to the lives of real people; what has become known as "human agency." We have seen further that another is to invoke the broader social, economic, and political structures that mold – even determine – health and health care provision. It is in the search for a middle ground that a third alternative to positivism has come to the fore in the past three decades. This is known as *structuration*, and is most closely identified with the British social theorist Anthony Giddens. Some geographers have drawn extensively on his theoretical work (see Cloke et al., 1991, for an overview).

While it is true to say that structuration theory has had less impact on the geography of health than, say, in historical geography, the discussion that follows does serve as a context for some examples of research that either explicitly or implicitly make use of it. Certainly, the work we examined in Chapter 1, on the lives of women with multiple sclerosis in Vancouver, may be thought of as situated within this kind of framework.

Structurationist Explanation

Structurationism recognizes the duality of structure and agency. That is to say, it acknowledges that structures shape social practices and actions, but that, in turn, such practices and actions can create and recreate social structures. One possible "language" in which this can be expressed is that of *time geography*, first outlined by the Swedish geographer Hägerstrand in the mid-1960s. Consider a time-geographic diagram (figure 2.5), in which members of an imaginary family engage in daily activities. For example, the female partner takes her young child to a pre-school centre, before going to work, stopping off to shop on her way back to collect the child, and then returning home. Her engagement in various activities consumes time at particular locations, while travel between such places also takes time. Although not shown in the diagram, her activity (and that of other family members) in particular settings brings her into contact

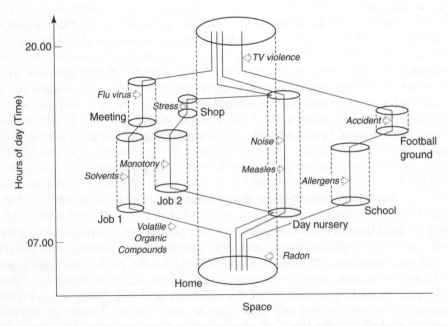

Figure 2.5 The time geography of an imaginary Swedish family (Source: Schærstrom (1996), *Pathogenic Paths? A Time Geographical Approach in Medical Geography* (Lund University Press, Lund), reproduced with kind permission of the author)

with others. While the purpose of this particular example is fundamentally epidemiological, used by the Swedish geographer Schærstrom (1996) to make the
point that time-varying exposure to various environmental problems and social
stresses has health consequences, the diagram also serves to give weight to the
interaction of structure and agency. Social structures require that particular
activities can only be carried out at particular times and in particular settings,
but, equally, such structures may themselves be transformed by social action.
For example, the opening times of clinics or health centers, together with workplace commitments, dictate when it might be possible to take the young child
to see a doctor or nurse; the structure of health care delivery constrains agency
(health-seeking behavior). On the other hand, the difficulties parents might have
in taking a child to be immunized (for example) and the consequent lack
of uptake of health care in particular places may ensure that the patterning
of health care resources in space is re-fashioned; in other words, agency may
transform structure. Time-geography has not gone uncriticized, notably by
feminist researchers (Rose, 1993: 19–40) who consider that it refers mostly to
masculine bodies and spaces.

Further Examples of Structurationist Approaches to the Geography of Health

Young (1996) has studied health care decision-making among women living in
contrasting suburban areas in Liverpool. Her particular concern is to consider
the "double burden" placed on women who are both part-time workers and
carers within the family. Many health policies assume that people are free to
make choices about lifestyle and usage of health services, and individuals get
"blamed" when they make unhealthy lifestyle decisions (for example, about diet
and smoking) or when they do not make "appropriate" use of health services
(for example, not attending for screening appointments). But, she argues, health
care decisions are *constrained* in a number of ways. One set of constraints is
explicitly geographical: the spatial reorganization of transport, shopping, and
health services that has produced greater concentration of such services and
increased reliance on access to private transport. Another set is fundamentally
economic: the fact that many women have low incomes and are also unpaid
workers at home. These and other constraints mean that women often neglect
their own health, or restrict their own food consumption, in order to protect
their children or partners. As one married mother of two explained, after receiving a welfare payment, she would "go straight to the chemist and get two
week's worth of baby food in, you know, the tins. And I thought well at least
if we've got nothing she's alright" (Young, 1996: 957). Human agency and
social structure both shape health behavior. Women, Young suggests, "must
make choices within the context of the economic and social network resources
available to them individually and through the family. Scheduling choices
must be judged against the time space frame of reference set by a woman's
own health status and the division of labor market and caring responsibilities
in her household" (Young, 1996: 955). Thus, as we saw in Dyck's work,

structuration theory provides a useful framework for feminist accounts of access to health care.

Dear and Wolch (1987) are more explicit than most in their use of structuration theory as they seek to explore the social processes involved in "deinstitutionalizing" the mentally ill; that is, taking them out of the psychiatric institution and into the community. These processes result in changes in spatial form, creating "landscapes of despair" or urban ghettoes of this and other "service-dependent" groups (including the dependent elderly, substance abusers, and ex-prisoners). Those discharged from institutions have gravitated towards inner urban areas, where they have found agencies willing to help and house them; as the numbers have swelled so too have more services appeared, leading to further in-migration as a self-perpetuating cycle of ghettoization emerges. For Dear and Wolch an analysis is required that recognizes both structure and agency in a reciprocal way. Long-lasting "macro-level" political and economic forces are at work, while these structures get translated into specific policies and programs by the state. Alongside these structures are the shorter-term routine behaviors of individual human agents who are involved in the social welfare system. As the authors put it, the "service-dependent ghetto has been created by skilled and knowledgeable actors (or agents) operating within a social context (or structure), which both limits and enables their actions" (Dear and Wolch 1987: 10). The reflexive or reciprocal relationship between social process and urban form arises in several ways. For example, inner-city improvement in recent years ("gentrification") has limited the possibilities for housing in the inner city; social processes influence spatial form. Yet the existence of the ghetto (urban form) shapes community awareness (social process) of the service-dependent community.

Post-structuralist Approaches to the Geography of Health

As we have seen, some of those engaged in the geography of health approach it from a positivist epidemiological stand-point, where measurable covariates are employed to shed light on disease incidence and key variables such as distance are used to explain disease spread. We shall encounter other examples later in the book. Others, we have noted, adopt interactionist approaches, exploring and interpreting the "authentic," lived experience of ill-health in particular places. Structuralists criticize both groups for their neglect of the broad social and economic forces, while structurationists acknowledge that both structure and agency matter.

More recently, however, some geographers, in common with other social scientists (and including some health researchers) have begun to engage with other theoretical developments, which may be labeled *post-structuralist*. In essence, these perspectives are concerned with how knowledge and experience are constructed in the context of power relations. This kind of perspective has illuminated work on health "risk," on representations of the body and of social groups, and on what it means to be a healthy citizen. It questions the rationalist assumptions on which much public health research is based (the so-called

Enlightenment or *modernist* tradition in which scientific "truth" reigns supreme). Since there are close links between public health and geography I think it is important to say something about post-structuralism. I should say at the outset that what some social scientists call post-structuralist others refer to as *postmodern*, but that – as with positivist and other approaches – there is no simple box into which we can place particular studies.

Further Examples of Post-structuralist Approaches to the Geography of Health

For writers such as Petersen and Lupton (1996) the "new public health" (which exhorts us to adopt healthy lifestyles – to eat well, exercise regularly, and cut down on smoking – as well as to play our part in creating healthy and sustainable environments) is a "modernist project." Having largely "solved" the problems of most infectious disease in the developed world, attention has shifted to the prevention of heart disease and cancers (and, latterly, AIDS), by promoting "good" health behavior. In doing so, control or power is exercised, not through repression but, as Michel Foucault has demonstrated, "through the creation of expert knowledges about human beings and societies, which serve to channel or constrain thinking and action" (Petersen and Lupton, 1996: xii) and, indeed, by self-governance. Petersen and Lupton see the new public health "as but the most recent of a series of regimes of power and knowledge that are oriented to the regulation and surveillance of individual bodies and the social body as a whole" (Petersen and Lupton, 1996: 3).

"Surveillance," of course, calls to mind the explicit monitoring of asylum populations explored by Philo (1989; see Chapter 1, above). But for these authors, the Foucauldian "gaze" is more subtle. Spatial controls of quarantines and "cordons sanitaires," and of isolation hospitals (and, to some extent, long-stay hospitals for those with mental illness and learning disabilities: the asylums studied by Philo), have given way to non-spatial controls implemented via legislation (such as the mandatory wearing of seat belts or controls on smoking in public spaces: Poland, 1998), inspection, and large-scale population surveys of those factors deemed to increase the risk of ill-health and disease. The direct "gaze" of the medical professional on the individual's body thus gives way to the social survey, whereby the health landscape of the population at large is "surveyed" and the "policing" of health is more subtle.

To illustrate this argument further Peterson and Lupton (1996: 120–45) consider the concept of the "healthy city," a public health project emerging from the World Health Organization for Europe (and since spawning other "healthy" social settings, including hospitals, schools, universities, and prisons). The basic aim was to bring together various organizations, associations, community leaders, and local citizens to promote and achieve better health for all in particular urban settings. The idea was that such cities would become models of good practice and that the idea would inspire other cities outside Europe to follow this lead. In Britain, an influential figure has been the Director of Public

Health in the North West Regional Office of the National Health Service, John Ashton (see, for example, Ashton 1992). The emphasis is on an holistic approach to health, with as much focus on housing quality, control of pollution, and "community spirit" as on measuring and monitoring of health status. The "sustainability" of urban life, the role of active citizenship, and the identification, monitoring, and reduction of environmental risk are all key features in the "healthy cities" movement. Cities are seen as ecosystems requiring management, but management that ideally involves active participation by all their citizens. If the city remains "unhealthy," rational, modern planning will ensure that the "organism" is restored to full health. Petersen and Lupton (1996: 127) are scathing in their attack on this fundamentally modernist project, which "obscures the power relations, uncertainties and ambiguities that underlie the development and implementation of policies, and conveys the impression that national, cultural and local differences, competing interests and inequitable access to resources are irrelevant to policy outcomes."

The emphasis on *difference* or *Otherness* is a key feature of poststructuralist accounts. These draw attention "to the fact that the assumed or constructed human subject of Western modernist discourse is an exclusive subject in that it is predominantly male, European, heterosexual, middle aged, and middle class" (Petersen and Lupton, 1996: 10). This rather sweeping generalization (somewhat ironic from postmodernists!) ignores the quite substantial body of public health research concerned with inequality, whether according to social class, gender, or ethnicity. But, having said this, let us consider a good example of how the "Other" has come under scrutiny from a health geographer, namely Craddock's study of the treatment of the Chinese community in San Francisco during the smallpox epidemics in the latter half of the nineteenth century (Craddock, 1995).

Craddock argues, in implicitly Foucauldian terms, that the analysis of disease in a geographical context "needs to include not just the visible configurations of spatial exclusion of the diseased, but also the many other ways the diseased are defined, disempowered, and controlled through metaphoric associations of place and affliction, inscriptions of contagious space, and the restructuring of purportedly diseased environments" (Craddock, 1995: 958). Smallpox probably diffused into San Francisco not with the Chinese but with other immigrants arriving from the eastern seaboard and the midwest. But given the need to find a scapegoat for the epidemics, and that the mechanisms by which the disease spread were poorly understood, the Chinese were blamed for bringing the disease into the city, not least by public health officials. As one of them observed in 1881, the Chinese, "coming in contact with our people generally as no other class of our inhabitants do. . . . are a constant source of danger to the health and prosperity of the entire community" (quoted in Craddock, 1995: 963). The appalling conditions in which they lived were seen as due to natural "depravity" rather than economic circumstances and the lack of jobs. Chinatown came almost to be equated with smallpox, and certainly stood as a metaphor for disease. It was a bounded area of the city that represented the "headquarters of disease." The Chinese "represented the most 'Other' of all others" (Craddock, 1995: 966) and the "threatening" urban space in which they were confined "was reproduced in the image of a threatening of disease" (Craddock, 1995: 967).

We saw in Chapter 1 how the "Panopticon" was considered as an architectural solution to the problem of monitoring "mad" people. Butchart (1996) has made explicit reference to the Panopticon in an industrial setting, exploring the treatment of black African mine workers in South Africa since the turn of the century. Here, however, the Panopticon is not only a built structure, but also a set of methods for clinical surveillance. These methods included, in the first few years of the century, the mapping of differential rates of illness and mortality among those migrating from different parts of southern Africa as a vehicle for shaping recruitment policy. Later, there was a more "differentiating and individualizing gaze" as medical officers sought to enquire into the incidence and pathology of tuberculosis among the mine workers. But the Panopticon was also realized in physical form, in the shape of the mine compounds. As contemporary writers described these in 1923, "dwelling huts are erected in long lines which radiate fanwise from a centre at which the compound offices are situated. The plan is devised to give maximum ease of supervision by the compound manager, who can survey the whole, or almost the whole, of his compound area whilst sitting at his office window" (quoted in Butchart, 1996: 194). Such structures were in operation until the 1970s, but are now replaced by "the computer, the bar code, and the swipe card, through which it is possible to constantly monitor the movement and location of every worker, both above and below ground" (Butchart, 1996: 195). Surveillance thus remains a tool of power whereby the worker can be controlled.

I want finally in this chapter to say something about *risk*. I do so in this section because some writers (notably the German sociologist Ulrich Beck: Beck, 1992) see a "risk society" emerging as a new form of modernity. Put simply, Beck means by this a society that is less concerned with visible risks that are produced by, or affect, the individual, or even uncontrollable risks arising from the natural environment; rather, one more concerned with large-scale, perhaps invisible, technological risk. Examples are the threat of nuclear contamination, of hazardous waste, or genetic modification of foodstuffs. These are new risks that have only emerged over the past 50 years. We no longer see only the benefits of modernization, but are aware of its hazards too. Beck's work, and that of others (for example, Wynne, 1996; Wakefield and Elliott, 2000), emphasizes that such risks are socially and culturally constructed, often by organizations and institutions. But since we are divorced from decisions that impact on our health and well-being we lack trust in the decision-makers or experts. As ordinary people we feel our own views must be examined and taken seriously, since we frequently have as much expertise as the "experts." These experts, with their quantitative estimates of risk (whether of pesticides, nuclear contamination, or food poisoning), dismiss lay beliefs as "irrational."

Beck's view has not gone unchallenged. Some writers (for example, Wildavsky: see Adams, 1995) argue that, in the western world, risks to health now are far less significant than at any time before. In the developed world we have a long life expectancy. We are unnecessarily and irrationally obsessed with risks to health, an obsession that is fueled by sensationalist media. We are too quick to panic. Most environmental and health risks are unproven, these authors argue, and we would do better to focus on the benefits to health brought about by modernization.

Concluding Remarks

This chapter has revealed the range and diversity of approaches to the geography of health. I think it is clear from what has been said – and I hope it will emerge from a reading of some of the subjects covered subsequently – that there are multiple interpretations of and possible explanations for particular health issues in geographical settings. For example, we have seen in this chapter that the spread of disease can be described statistically, or, alternatively, explained in terms of the impacts of colonial structures. We shall see in Chapter 4 the richness of variety in accounting for health inequalities; for some, the effects of place and space are captured by positivist epidemiological approaches, while others seek structuralist, class-based explanations, and yet others give more credence to ordinary people's accounts of such inequalities. We shall see in Chapter 7 that an understanding of the impacts of air pollution on health can be understood by adopting a positivist "risk factor" approach, or a social interactionist account that gives weight to "lay" or popular views of the association, or a structuralist interpretation that the world industrial-economic system is the root cause of the problem.

I think it appropriate as well to end on a cautionary note: "Any attempt to categorize the theoretical approaches taken in medical geography is surely doomed to be flawed and partial, to illuminate some aspects of the intellectual landscape while obscuring others, and in doing so to be just one possible way of telling the story among many" (Philo, 1996: 36). I do not wish to put an intellectual straightjacket onto the work done by health geographers and allied workers. The "landscape" is complex and does not require bulldozing. It does require a map, however, and this is what I have endeavored to provide. But just as all maps are models of the territory they purport to describe, so is my own simply one, hopefully reasonable, sketch of the recent historical and contemporary terrain.

FURTHER READING

I have followed others, in both the geographical and health research arenas, in my threefold classification of work as "positivist," "social interactionist," and "structuralist." See, for example, Aggleton (1990) among the latter and Litva and Eyles (1995) among the former. However, other important bodies of social theory need to be addressed, and I have sought to give due weight to "structurationist" and other "post-structuralist" accounts. Curtis and Taket (1996) adopt a similar approach, though with a somewhat different classificatory scheme.

For a wonderfully clear justification for taking philosophy seriously, as well as a succinct journey through some of the debates I recommend that the geographer begins with:

Graham, E. (1997) "Philosophies underlying human geography research," in Flowerdew, R. and Martin, D., eds. *Methods in Human Geography* (Addison Wesley Longman, Harlow), before moving on to Cloke et al. (1991). The latter gives an extended treatment of "humanistic" approaches (what I have termed social interactionist), as well as structuralist (Marxist), structurationist, and postmodern approaches.

On broadly "positivist" styles of the geography of health see: Cliff and Haggett (1988) and the wide-ranging analyses of disease records for world cities in the late-nineteenth and early-twentieth centuries (Cliff et al., 1998) as well as for island populations (Cliff et al., 2000). An excellent pedagogic account of classical spatial diffusion models is given in Thomas (1992).

Several of the essays in the collection edited by Butler and Parr (1999) draw on a social interactionist perspective; moreover, the editors' introduction gives a very good insight into contemporary geographies of health and disability.

For a very concise summary of the impact of colonialism on the health of native populations see the introductory editorial in Arnold (1988), while among a number of comprehensive syntheses I have drawn on Watts (1997). A series of detailed case studies is presented in the collection edited by Hartwig and Patterson (1978). Detailed Marxist-based accounts of the development of health care and the impact of colonialism on disease outbreaks in Tanzania (Tanganyika) are given in Ferguson (1979) and Turshen (1984).

On feminist scholarship see Rose (1993), but also the recent collection of essays edited by Dyck and others (2001).

The classic text by Jones and Moon (1987) has a great deal to say about different perspectives on the geography of health. It is particularly valuable on positivist and structuralist (referred to there as "materialist") accounts. Although some of the examples are a little dated, the book should be read from cover to cover.

Rose (1993) reflects critically on structurationist approaches, particularly inasmuch as they have tended to ignore women.

Butchart's work on the African mining compound was taken as an example of a Foucauldian perspective on health; see his book, Butchart (1998), for an extended treatment, looking, for example, at missionary medicine too. An important recent paper on AIDS, with particular reference to Malawi, and drawing on post-structuralist perspectives, is by Craddock (2000).

For good introductions to risk, see Adams (1995), though the classic text by Beck (1992) merits a careful reading. See also the collection of essays edited by Heyman (1998). The book by Petersen and Lupton (1996) has much to say about risk, and casts a wonderfully critical, and implicitly geographical, post-structuralist eye on public health.

CHAPTER 3

METHOD AND TECHNIQUE IN THE GEOGRAPHY OF HEALTH

Having set out some broad approaches to studying the geography of health I want now to consider the variety of methods and techniques that can be brought to bear in empirical studies. We have already encountered some of these in the examples used in previous chapters. For example, studies of disease incidence and spatial diffusion tend to use cartographic and statistical methods, while studies of how people with ill-health or disability cope with their environments call usually for in-depth interviews with small numbers of people.

The present chapter is divided into two substantial sections. The first examines what I shall call the "mapping" of health. Here, we consider not only techniques for visualizing health outcomes but also some of the tools of quantitative analysis that geographers and others use when studying the spatial distribution of disease and illness. My aim is not to produce experts in quantitative spatial analysis but rather to try to get across the flavor and range of methods that are used; readers will have to look at the source material suggested at the end of the chapter for details.

The second section examines a range of methods that are essentially qualitative; here, the aim is less on measurement and more on the interpretation and understanding of ill-health, disease and disability in the context of place. Again, this material is not sufficient to produce experts in qualitative data analysis, but as with the first section I hope it is enough to allow the reader to appreciate some empirical work considered in this book and elsewhere.

I should also say, by way of introduction, that it is a mistake to equate quantitative methods with positivist accounts, and to assume that other approaches to the geography of health simply require qualitative methods. In practice, this will usually be the case. But what matters most is simply whether the methods and techniques can be considered to have shed any light on the problem being studied.

"Mapping" the Geography of Health: Quantitative Approaches

Quantitative spatial data analysis typically involves one or more of three tasks. Firstly, we will usually want to draw a map of our data, in order to see whether there seems to be any spatial patterning or perhaps visual evidence of an association with social or environmental factors. I will refer to this as *visualization*. Second, we will almost always want to use both graphical and statistical methods to explore the data; this involves a more rigorous search for spatial pattern and possible associations with other variables and is referred to as *exploratory spatial data analysis*. The dividing line between "visualization" and "exploration" is a fine one, since both rely heavily on graphical methods. Third, we may wish to specify a more formal statistical model of our data; typically, such *modeling* will involve testing a hypothesis (for example, that the incidence of a disease is higher around a suspected source of air pollution). In the geography of health, such models will usually involve a "spatial" variable (such as distance from a pollution source, or perhaps a simple indicator to distinguish one area from another). Again, the division between exploration and modeling is fluid, and this three-fold separation is one of convenience. In practice, analysts often use all three simultaneously, using modern statistical software in order to move interactively from a map, to another picture, to the fitting of a model, and then mapping the results (Bailey and Gatrell, 1995).

Visualization

In order to draw a map of health data we need some form of spatial referencing. If data have been collected for a set of individuals we might have some form of postal code (not usually a complete address, in order to protect confidentiality) that can be linked to a "point" on the ground. As noted in Chapter 1, a UK postcode "maps onto" an Ordnance Survey grid reference, for example. We could then map the set of individuals who are ill or have been registered on a database. In practise, this will not be very useful, since it will tend to reflect the distribution of population as a whole. Instead, as we shall see in the next section, we need some data on a "control" population in order to make comparisons. The simple visualization, as a dot distribution map, of "point" data in epidemiology is often not informative.

If, however, we aggregate individuals to a set of areal units, we can draw more useful maps. Commonly, data come in this form anyway. For example, Betemps and Buncher (1993) list data on the distribution of motor neurone disease by state in the USA; the Leukaemia Research Fund Atlas (Leukaemia Research Fund, 1990) gives data on the incidence of childhood leukaemia by county in Britain; while many web-based resources, some listed in the Appendix (page 256), provide access to such data. Unless one is working with health professionals (or those controlling access to individual health records) the geographer will most probably be using area data.

In practise, geographers rarely map observed counts of cases by area, since these too would simply reflect population distribution or the age structure of that population. Even mapping crude rates (for example, number of cases of heart disease per 10,000 resident people) is usually of little value, since it will not have allowed for the age structure of the population; older people are more

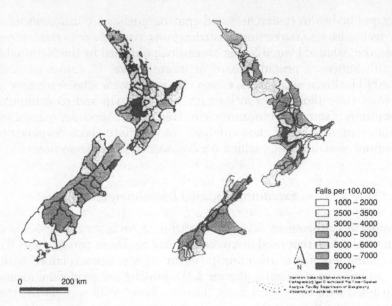

Falls per 100,000
☐ 1000 – 2000
☐ 2500 – 3500
☐ 3000 – 4000
▩ 4000 – 5000
☐ 5000 – 6000
▩ 6000 – 7000
■ 7000+

0   200 km

Figure 3.1 Mapping morbidity in New Zealand: "conventional" map and a cartogram (courtesy of Daniel Exeter)

likely to suffer from heart disease. Consequently, some form of standardization is adopted before data are mapped (see box 1.1 page 6).

There are many issues, of both a cartographic and a statistical nature, raised by this kind of mapping. Cartographers have long wrestled with ways of defining class intervals for these so-called *choropleth* maps; how many classes should we use, what should the class intervals be, and what shading or color scheme should be used? These are not trivial issues, since they affect how the map is "read" or perceived (both Cliff and Haggett, 1988, and Monmonier, 1996, have a good discussion of these and related visualization issues). A further difficulty is that the areas themselves are of variable physical and population size. As a result, when we shade the areas we may give too much "visual weight" to quite large, often rural, areas that may be sparsely populated. One simple, and sensible, cartographic solution is to shade only those areas above a threshold population size, and to place a small shaded symbol in the less densely populated areas. Another is to draw a *cartogram* or *isodemographic* map, which transforms the areal units according to population at risk, such that rural areas appear much smaller than on a conventional map while urban areas "expand." This idea has a long history in disease mapping, though it has not been widely used, partly because of the tedium of performing the map transformation and partly because the resulting maps, while sensible, are difficult for lay readers to grasp. However, Dorling (1995) has popularized them in his "New Social Atlas of Britain," which includes numerous maps of health outcomes. An example from New Zealand is shown in figure 3.1.

There are many examples of more conventional published atlases of disease and ill-health, showing spatial variation at a number of different scales. More recently, public health researchers and epidemiologists are using modern information technology to make these visualizations available over the Internet. For example, the Atlas of United States Mortality, published by the National Center for Health Statistics, provides maps of mortality for 18 causes of death, for nearly 800 Health Service Areas. Given that many such atlases are now in electronic form they allow users to interact with the map and to manipulate the data themselves; some reference sites are given in the Appendix (page 256). The possibilities of interactive data analysis lead us into data "exploration" or "exploratory visualization," which we consider in the next section.

Exploratory Spatial Data Analysis

In addition to problems in mapping health data for a set of areal units, there are statistical issues that need discussing. I shall highlight two of these. The first of these is the *modifiable areal unit problem*. This is best explained with reference to an imaginary example (figure 3.2), showing the residential locations of children born with congenital heart disease. As we shall see shortly, there are ways of exploring this using the point data themselves, but if we wish to analyze how the data are distributed by area we will get very different results depending upon how we configure the areal units. A configuration based on administrative units (such as electoral wards in Britain, counties in the United States, communes or municipalities in many European countries) will give different estimates of disease incidence than one based on different (and essentially arbitrary) zones or ones that show "bands" of distance from a suspected source of pollution (figure 3.2). We say that the areal units are "modifiable." The British geographer Stan Openshaw has been at the forefront of identifying and examining this problem (Openshaw, 1983).

The second issue I wish to discuss relates to the numbers of cases used to construct estimates of disease incidence, particularly SMRs. To some extent this is a problem of scale, since as our areal units become larger the numbers of cases will grow. For example, we will observe more lung cancers at the state rather than county level in the US, and more cases at the provincial level in Italy than at the municipality level. When mapping and analyzing data for well-populated areal units we can put a considerable degree of confidence in our estimates of the SMRs, much more so than at a fine spatial scale, where the counts may be very small (perhaps one case here, two or three there, none in some areas, and so on). At this finer scale we say that estimates of disease risk are "unstable," since the addition or subtraction of only a single case can greatly alter the estimate. For example, if over a three-year period we expect two deaths in one area, and observe three, we have an SMR of 150. But if we observed none, or one, or two, or five the SMR would change to 0, 50, 100, and 250. This is the so-called *small number problem*.

There are a number of possible solutions. One is to extend the data collection period to cover further years. A second is to work at a coarser scale (though this is of little help if we wish to identify local disease "hotspots"). A third strategy is to place confidence intervals around the estimate of the SMR (so we can

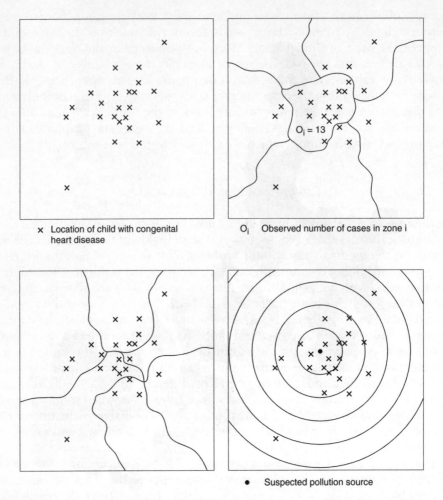

Figure 3.2 The modifiable areal unit problem: assigning point data to different sets of areal units

tell whether a rate of 130 is significantly different from 100, for example); this is widely used in disease mapping. A fourth alternative is to use *probability mapping*, where we map the probability of observing a count that is more extreme or unusual than that actually observed. Technically (Bailey and Gatrell, 1995: 302) we use the Poisson distribution to calculate these probabilities. Low probabilities (say, less than .05) suggest that an area's rate is significantly high if the observed value is greater than the expected, or significantly low if the converse is true. A map of Poisson probabilities for child mortality in the city of Auckland, New Zealand (Marshall, 1991) highlights some such areas (figure 3.3).

The difficulty with probability mapping, however, is that it does not allow for the fact that incidence in nearby areas may be correlated (the Poisson dis-

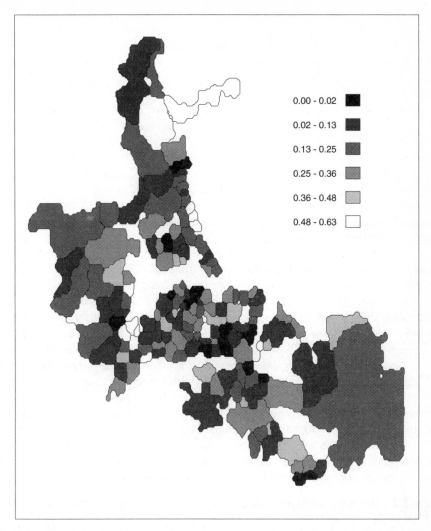

0.00 - 0.02	■
0.02 - 0.13	■
0.13 - 0.25	■
0.25 - 0.36	■
0.36 - 0.48	▨
0.48 - 0.63	□

Figure 3.3 Poisson probabilities of child mortality in Auckland, New Zealand (Source: Bailey and Gatrell, 1995)

tribution assumes that the count in one area is independent of that in other areas), while it tends also to highlight those areas with larger populations. SMRs, by contrast, tend to pick out areas with smaller populations. More recently, therefore, statisticians have suggested a compromise, using what are called *shrinkage*, or more formally, *empirical Bayes* estimators of disease risk. Essentially, empirical Bayes estimation does the following. Where the estimate of disease risk is quite reliable or stable, because it is based on large numbers of cases and a sizeable population at risk, it is left untouched. However, if it is unreliable, because it is based on small numbers and a sparsely populated areal unit, we "shrink" it towards the average or mean rate for the study area as a whole; this is a way of professing our "disbelief" in the observed rate. More sophisticated versions of this idea let the rate in a small area "shrink" to a local or neighborhood mean (using only areas adjacent to the one in question) instead of the mean for the entire study area.

To illustrate the idea, consider the problem of mapping suicide rates in England and Wales (Saunderson and Langford, 1996). During the 1980s the rate among young men (aged 20–24) increased by 71 percent, clearly a matter of great social concern. But is this worrying trend spatially uniform, or are there particular areas of high and low risk? The data analyzed by the authors relate to incidence across 401 districts between 1989 and 1992, in different age groups, and among women as well as men. But if we focus on men aged 25–44 years, where the incidence of suicide (17.2 per 100,000) is over four times that of women (3.9 per 100,000), the SMRs range from zero to over 250. Geographically (figure 3.4) there are pockets of high-risk districts in all parts of the country. The empirical Bayes estimates suggest that much of the variability is because of small numbers of suicides in some districts and the range of estimated risks is therefore reduced. When mapped (figure 3.5) the estimates confirm that there are areas of high risk in parts of Devon and Sussex (southwest and south-east England), parts of north-west England, and in north Wales and Lincolnshire. These are largely rural areas. In addition, there are some high risk areas around major cities (Manchester, for example). Saunderson and Langford suggest that the rural "excess" may be explained by unemployment in agriculture (especially among those employed in hill-farming and dairy production) and by social isolation. Given the crisis in British farming during the rest of the 1990s, caused by low prices and loss of confidence among consumers in the wake of the BSE ("mad cow disease") fiasco (box 3.1), we can speculate that this rural clustering will have worsened. Unemployment, social isolation, and rates of mental illness are also factors likely to explain high rates of suicide in some urban areas. Of course, to shed real explanatory light on these variations requires more in-depth work, perhaps along the qualitative lines we consider shortly.

Assuming we have constructed a map of disease incidence, what might we do with it? We have already suggested that the map may show evidence of spatial patterning. Do areas of high incidence "cluster" together, for example, or is the spatial arrangement of disease rates essentially random? Non-randomness might suggest clues about disease causation. Such spatial patterning is referred to as *spatial autocorrelation*, and it is a fundamental concept in geography. Using an analogy with ordinary correlation analysis in basic statistics, we speak of positive autocorrelation where similar values tend to occupy adjacent locations on the map, and negative autocorrelation where high values tend to be located next to low ones. If the arrangement is completely random we refer to an absence of spatial autocorrelation. Techniques for measuring autocorrelation using a test statistic are discussed in Bailey and Gatrell (1995, Chapter 7).

There are good reasons why we should expect spatial autocorrelation to be a very common feature of geographical data in general and health data in particular (Haining, 1990: 24–6). First, as we saw briefly in Chapter 2, the process of spatial diffusion or disease spread means that the incidence of measles or

BOX 3.1   *Mad cows and risks to human health*

Bovine spongiform encephalopathy (BSE, popularly known as "mad cow disease") is a progressive, fatal disease of the nervous system in cattle, manifesting itself in the stumbling gait of the animals. Its human equivalent is known as Creutzfeldt-Jakob disease (CJD), the annual incidence of which is roughly one per million people. BSE was first discovered among British cattle in the mid-1980s, with over 160,000 cases reported by late 1996. The theory is that BSE spread because of the consumption of infected high-protein animal feed; the offal from cows and sheep being used to create a food product for other cows.

CJD tends to affect the elderly, but in the mid-1990s cases began to emerge in much younger adults; among these, the disease progressed differently and was characterized by larger aggregations of the prion protein (thought to be the infective agent) in the brain tissue. This was a new variant of CJD (nvCJD, now called vCJD). An advisory committee (UK Spongiform Encephalopathy Advisory Group, SEAC) reported in 1996 that the cases were likely to have resulted from consumption of material from BSE-infected cattle before the 1990s. As of August 2000, 79 confirmed and probable cases of vCJD had arisen in England and Wales.

Cousens and his colleagues (1999) mapped the distribution of cases as of August 1998 with a view to testing the hypothesis that they were clustered near animal rendering plants. The hypothesis was not confirmed, although there were more cases in the county of Kent, south-east England, than chance would dictate. Conceivably, this might be related to consumption of locally produced, and contaminated, beef.

More recently, a "cluster" of cases of vCJD has come to light in the village of Queniborough, Leicestershire. Here, four people who have lived in the village have been diagnosed with the disease. Given the rarity of the disease this is a highly unusual local geographical aggregation, and public health specialists investigating it are optimistic that it may lead to final confirmation of the link, or otherwise, to the consumption of contaminated beef. In the meantime it is difficult to predict the number of future cases of vCJD. Estimates vary from a few dozen to many hundreds of thousands.

For further background to the BSE and vCJD problems see Powell and Leiss (1997). For an initial examination of the geography of vCJD in Britain see the *Lancet* (1999, 353: 18–21 and 1357–9).

influenza, for example, in one area is likely to affect directly the incidence in nearby areas. Second, there will often be broad regional processes, whether environmental or social, which ensure that disease incidence in one small area is similar to that in neighboring areas. For example, if there is widespread regional pollution it is likely that areas within this wider region will all tend to have high rates of respiratory disease. Or, if this wider region has quite uniform

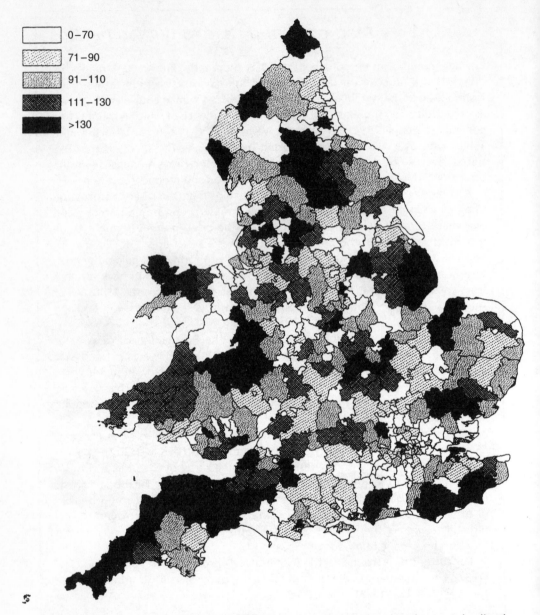

Figure 3.4 Suicide among men aged 25–44 years in England and Wales: standardized mortality ratios (courtesy of Ian Langford)

levels of poverty we can expect rates of some chronic diseases to be high in adjacent small areas that make up that region, since poverty is closely linked to many such diseases. Geographers and statisticians have collaborated to devise statistical measures of spatial autocorrelation, and the use of an autocorrelation statistic (Cliff and Haggett, 1988: 33–5; Bailey and Gatrell,

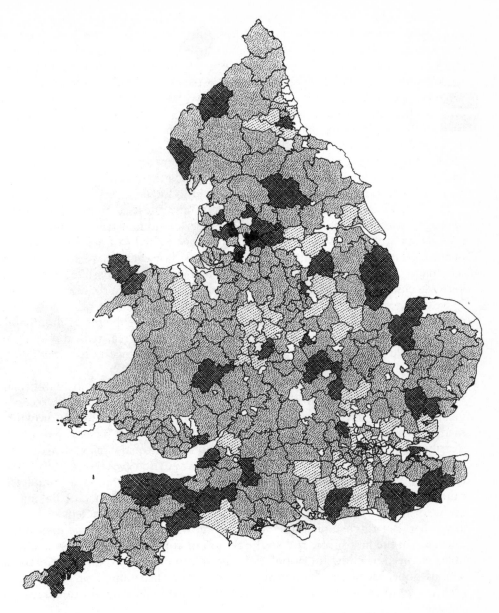

Figure 3.5 Suicide among men aged 25–44 years in England and Wales: empirical Bayes estimates (courtesy of Ian Langford)

1995: 269–74) in characterizing health data is now widespread. For instance, Walter et al. (1994) detected spatial autocorrelation in the incidence of some cancers in Ontario, Canada, including stomach and lung cancer for both men and women.

To summarize a whole map pattern by a single number is a little crude, and this has led others to suggest more "local" measures of spatial autocorrelation or spatial association. In a health context, these look at the association between a disease rate in one location and rates in neighbouring locations, up to a specified distance. This helps identify disease "hot-spots." Details of the method are given in Getis and Ord (1992) and applications to health data are presented by these authors in that paper and elsewhere (Getis and Ord, 1999). For example, they study the clustering of AIDS cases in the area surrounding San Francisco, between 1989 and 1994. Results suggest that there is a strong tendency for clustering of high rates among the seven counties within about 40 miles (64 km) of San Francisco. Rigby and Gatrell (2000) have used the method to look for clustering of breast cancer in north-west Lancashire, England.

We have already drawn attention to the arbitrariness of areal units, though in many cases, as when presented with data for administrative units whose boundaries are fixed, we can do little about this. Systems of areal units can be defined using geographical information systems, however, as we see later (box 3.3, page 73). For example, if we wish to explore the possible association between childhood respiratory disease and proximity to main roads (see Chapter 7) we can create "buffer zones" around the roads and count the number of cases of ill-health within and outside such zones. The same problems of modifiability arise as before (see figure 3.2), since there will be no real reason why these zones should be 25 metres wide, or 50, or 100; the choice is to some extent arbitrary.

The arbitrariness of areal units, coupled with the increasing availability, in the developed world, of coordinate data derived from individual health records, means that a number of geographers and epidemiologists have been drawn to methods based on spatial point process analysis. Here, we have a map of the distribution of observed health "events" (points in space) and, as with area data, interest centers on whether the map could have arisen by chance. More formally, we might ask whether the distribution is the outcome of a completely random process, such that any sub-region has an equal chance of having an event located there and the occurrence of one event is independent of that of any other. But this is unhelpful, since we know that population is distributed unevenly in the first place, and so the notion of complete randomness is untenable. We need to allow for population distribution in our exploratory analysis. This is done, ideally, by comparing our observed distribution of point events ("cases") with a set of "controls" (similar individuals who do not have the disease).

Consider an example relating to the incidence of congenital heart malformations among children born in north Lancashire and south Cumbria, north-west England, between 1985 and 1994. These data (figure 3.6a), comprising the home addresses of 138 cases, can be matched to a set of 276 controls, healthy infants who are the immediately preceding and following births from birth registers (figure 3.6b). How might we use this information in an exploratory way?

One question is to ask whether the cases are "clustered;" of course, they are, in the towns and villages! So we ask if there is clustering over and above that

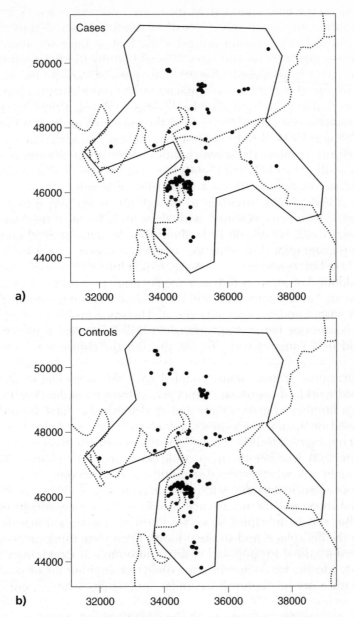

Figure 3.6 Congenital heart malformations and a sample of healthy infants in north Lancashire and south Cumbria (1985–1994)

which we might expect based on "population distribution" (which we take to be sensibly represented by the sample of healthy infants). There are different ways in which we can answer this question. One uses a statistical function known as the *K-function* (Bailey and Gatrell, 1995). In a sense, this is a description of the level of spatial autocorrelation in point data; does the occurrence of a point event at any location make it more, or less, likely that we will find a point nearby, or is there no such correlation? Graphically, one can think of estimating a K-function by superimposing a finely spaced set of concentric circles over each event, and then calculating the mean number of events within a particular distance of any arbitrarily chosen event. In practice, an allowance is made for events close to the study area boundary, while a transformation of the K-function is sometimes used for graphical display (Bailey and Gatrell, 1995: 90–5). The (transformed) K-function for the cases is shown in figure 3.7a, and the fact that it is positive over all distances is evidence of clustering, as expected; there are many more points close together than we would expect by chance. But if we estimate a K-function for the controls we can then ask if the two estimated functions are significantly different. Since the K-function for the controls (figure 3.7b) is very similar to that for the cases the difference between the two is negligible and well within the limits of no significant difference (figure 3.7c). This suggests that the two sets of point events, cases and controls, could have come from the same population; there is no clustering of congenital malformations once we have allowed for "natural" population clustering.

This technique assesses whether or not there is clustering in the study region. Another, different, question is to ask if there are *clusters* and, if so, where these might be located. This problem can be tackled using a technique called *kernel estimation*. This seeks to estimate the density or "intensity" of the spatial point process (Bailey and Gatrell, 1995: 84–8; 126–8). Again, a graphical analogy is perhaps more helpful than statistical detail (figure 3.8). Here, we imagine placing a "hat" (a kernel function), of fixed shape and radius (known as the bandwidth) over all locations on the map, and at each location estimating the local density. We do not simply count the number of point events within the radius; rather, this is weighted according to distance from the location at which density is being estimated. Again, allowances or corrections are made for edge effects. More complex versions of this idea allow the bandwidth to vary, being narrower in sub-regions of high density and wider elsewhere. As before, estimating density for cases of disease will simply identify clusters of population. But using controls, we can ask whether clusters of cases occur more, or less, than we would expect given the distribution of the control population. This is done by estimating the ratio of the two kernel estimates, which should be constant across the map if there are no clusters. Again, because of sampling fluctuation there will always be some disparity between the two kernel estimates, but we can use statistical methods to assess the significance of "peaks" on the surface of cases; do these cluster in one sub-region more than we would expect on the basis of chance? Analysis of our malformation data (figure 3.9) suggests that there are no genuine clusters of cases once we allow for the distribution of healthy infants. As with all statistical techniques, we are more likely to detect differences if sample size increases; with only 138 cases our sample is too small to detect real differences, even though there is visual evidence of a possible cluster around the small town of Windermere.

These are simply two techniques among many that are used to look for "clustering" or "clusters." Others may be found in the suggested reading given later, along with further details of the methods considered above. One well-known

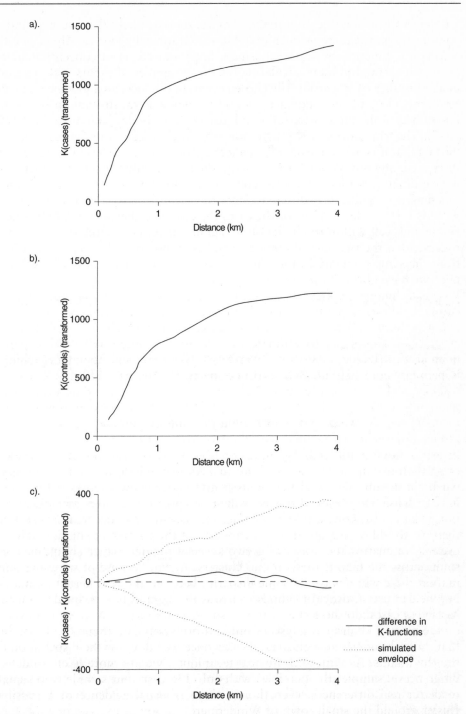

Figure 3.7 K-functions for congenital malformations and healthy infants, and the difference between the two

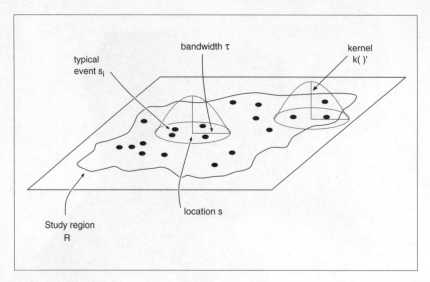

Figure 3.8 Kernel function for a spatial point pattern (Source: Bailey and Gatrell, 1995)

method for detecting clusters is Openshaw's "Geographical Analysis Machine" (Openshaw et al., 1988), described in other texts (Thomas, 1992).

Modeling Health Data in a Spatial Setting

In exploratory spatial analysis we are not seeking to test explicit hypotheses, other than perhaps trying to assess departures from randomness. For example, when considering the incidence of congenital heart malformations we did not have in mind any notion that there might be clusters around a possible pollution source. This contrasts with situations where we do believe, *a priori*, that there might be a raised risk of disease near such a source. The best-known example of this kind of "focused" study is the substantial research effort devoted to the suspected elevated risk of childhood leukaemia around plants processing nuclear material. Controversy surrounding the nuclear reprocessing plant at Sellafield on the Cumbrian coast of north west England (see box 3.2, page 66) has motivated these studies.

Again, there is a range of point-based methods that can be put to use on this kind of problem. One (Diggle and Rowlingson, 1994) allows us to test whether the incidence of disease is related to one or more possible sources of pollution, while controlling for the possible effects of other variables allowed for at the individual or household level (such as smoking behavior, or damp in the home). Essentially, this approach takes a set of point locations and asks: how likely is it that any given point is a case rather than a control, given its distance from one or more suspected pollution sources and any additional information we may have about the individual in question? Some researchers have, for example,

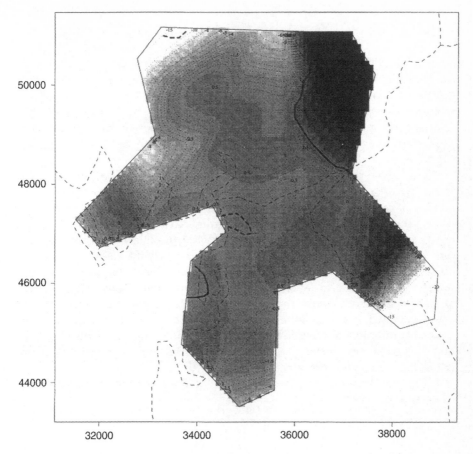

Figure 3.9 Ratio of kernel estimates for cases of congenital malformations and healthy infants

asked if proximity to sites disposing of hazardous waste, whether via landfill (Dolk et al., 1998) or incineration (Diggle et al., 1990; Elliott et al., 1992a), is a risk factor for congenital malformations or cancers of the respiratory system, while others have asked if there are clusters of breast cancer on a Long Island community in New York (Timander and McLafferty, 1998).

Modeling of area data is much more common, since we invariably have census-based data to use as potential explanatory variables (also often known as covariates). The standard tool to use here is regression analysis (box 1.3, page 16), which seeks to account for variation in a dependent variable (such as disease incidence) in terms of the chosen covariates. Lovett and Gatrell (1988) sought to account for geographical variation in rates of spina bifida (a congenital malformation of the central nervous system) among health authorities in England and Wales. Statistically significant predictors were: percentage of female residents born in the Caribbean (which was inversely related to spina bifida); percentage unemployed; percentage of the workforce employed in agriculture (both

BOX 3.2   *The Sellafield leukaemia "cluster"*

Television publicity was given in 1983 to an unusually large number of cases of childhood leukaemia near the Sellafield nuclear reprocessing plant in West Cumbria, England. Numerous epidemiological studies (including those by geographers: see Openshaw et al., 1988) have confirmed the existence of this "cluster" and indicated that it is restricted to children born in the nearby village of Seascale (6 cases of leukaemia observed in such children, compared with only 0.6 expected).

The Sellafield problem has motivated a considerable volume of sophisticated technical work on the detection of disease clusters. The explanation of the excess number of cases of childhood leukaemia near Sellafield is, however, controversial. For some, it arises as a result of direct exposure to contamination on nearby beaches or via consumption of seafood and locally produced foodstuffs. Some, such as Gardner and his colleagues (1990) suggest that the raised incidence is associated instead with fathers' employment at the Sellafield plant. The high radiation doses given to fathers, before the conception of any offspring, has resulted in leukaemia among their children (a result of a mutation in the sperm). Others contest this as a causal mechanism. In particular, the epidemiologist Leo Kinlen has offered a competing hypothesis based on unusual patterns of migration and population mixing; his work, arguing the role of infectious agents rather than radiation, is considered in Chapter 6.

For further discussion see Gardner (1992) and the collection of papers edited by Beral et al. (1993).

positively associated with spina bifida); and, most significantly, an indicator ("dummy") variable that denoted whether or not the health authorities routinely screened pregnant women for the disease. As with all regression models in health research that use spatial units for their analysis, differences between observed and expected incidence can be mapped and the results suggest parts of the country where, in particular, incidence that is higher than expected merits particular attention.

This example serves to make a number of points. First, there is nothing particularly "spatial" about the analysis. All the health authorities could be randomly located anywhere on the map and the regression results would be identical! "Place" enters into the analysis merely as a container for the disease cases. More complicated models of disease incidence allow, as covariates, spatially weighted averages of the data in surrounding areas to be used (see, for example, López-Abente, 1998). Second, census data are used frequently in studies of this form simply because the required data do not exist at an individual level. We can hardly conclude from the spina bifida study, for example, that a woman born outside the Caribbean, and whose partner or herself is unemployed, is more likely to give birth to a spina bifida baby. The absurdity of this argument is born out by the fact that employment in agriculture is also a variable in the analysis; it is impossible to be both unemployed and to work in agriculture! We would need to have individual-level data, and data for a set of controls, in order to test these kinds of hypothesis.

It does not follow from this that we *either* conduct analysis on data that are unavoidably aggregated to a set of areal units, *or* that we conduct analysis on individual-level data. Indeed, arguably the most significant contribution made by geographers to quantitative health research has been to demonstrate that data for *both* places *and* individuals can be brought together in order to shed light on health outcomes. This body of work is known as *multi-level modeling*, and it merits consideration here. Such modeling can get quite complicated, and all I do here is sketch some basic ideas, again relying on graphical rather than mathematical formulations.

Suppose we wish to explore the relationship between smoking behavior and age. We have data for a large number of individuals living in several towns; we know their age and how many cigarettes they smoke on an average day, as well as their place (town) of residence. Pooling all individuals and fitting a simple regression model would yield a regression line with a fixed intercept and slope (figure 3.10a). But individual smoking behavior might vary from place to place; perhaps there is the same overall relationship between smoking and age, but each town has its own intercept, denoting its varying average cigarette consumption (figure 3.10b); here, the intercepts are said to be "random" (meaning, drawn from a probability distribution). In this case, some places simply have higher overall levels of smoking, at all ages. Or, the relationship between smoking and age might itself vary from place to place. In some places there might be no relationship (the slope is horizontal); in others it might be quite steep (as age increases, cigarette consumption rises rapidly); in yet others there might be an inverse relationship (figure 3.10c). The slopes and intercepts are both "random." There is, then, a distribution (over the set of places or towns) of possible intercepts and slopes, the relationships between which can be explored. The main point is that these place-varying behaviors are masked by a single-level analysis.

In this example, we are using two levels (individuals and towns), but there might be three or more levels. For instance, children with asthma might be affected by age (individual level), by the presence of smoke or damp in the home (a higher-level effect), and by the volume of traffic along their street (a third level). There may be interactions across levels too; for example, the risk of lung cancer will be elevated in those who smoke heavily (an effect at the individual level) *and* who live in areas where there is substantial air pollution (an effect at a higher level). We need, then, to add to our analysis a set of variables that are measured at the higher levels.

Kelvyn Jones and his colleagues have applied these ideas in a number of different contexts; let us consider one of their empirical studies (Jones and Duncan, 1995). This relates to respiratory ill-health in England, taken from the Health and Lifestyle Survey (Cox et al., 1993). Data are recorded for 9003 individuals, living within 396 small areas (electoral wards). As we shall see in Chapter 4, it is possible to create a deprivation score for such areas. Variables measured at the individual level include age, height, gender, smoking and eating behavior, alcohol consumption, exercise, income, tenure, social class, and employment status. The problem is to account for variation in respiratory health ("forced

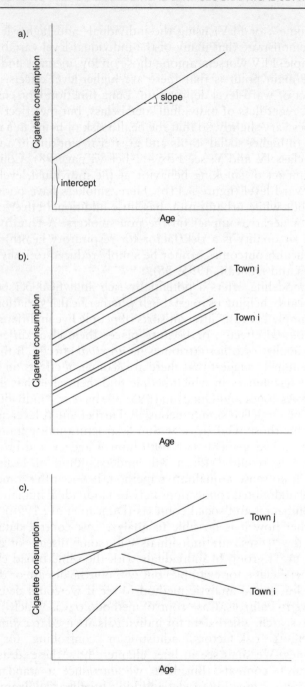

Figure 3.10   Multi-level modeling of hypothetical data

expiratory volume", or FEV) using the individual- and higher-level variables. The authors demonstrate that many of the individual-level variables are significant; for example, FEV worsens among those on low incomes and of low social status. But the main point is that there are higher-level effects. Figure 3.11a shows the effect of ward-level deprivation. Lung function worsens in the more deprived areas, regardless of individual social class; but the effect of ward-level deprivation does vary slightly, so that the health risk of living in a deprived area is less if one is of higher social status and greater if working in a manual occupation (social class IV and V: see box 4.2 below, page 98). Consider too the simultaneous impact of smoking behavior at the individual level, and urbanization at the ward level (figure 3.11b). Here, smokers have poorer FEV than non-smokers; but while urbanization has little additional effect on smokers it seems to have a negative impact on the non-smokers. A tentative conclusion would be that air quality is a risk factor for respiratory health. These results indicate that "health outcomes cannot be simply reduced to individual characteristics" (Jones and Duncan, 1995: 38).

Multi-level modeling offers a potentially rich analytical perspective on the geography of health, helping us to establish whether health variations from place to place arise simply because different sorts of people live in different places (so-called *compositional* effects), or because places themselves differ in terms of environmental quality or other attributes (*contextual* effects). If the latter is the case, we can genuinely suggest that there are *ecological* effects on health, in the sense that the environment in which people find themselves has direct impacts on their health. As Jones and Duncan (1995: 30) have it, "individuals and their ecologies must be modelled simultaneously." Further, multi-level modeling is an attractive tool for those who work within a structurationist framework, since it suggests that an engagement with both human agency and broader (higher-level) structures is required for a full understanding of health problems. Although it is in no sense a qualitative method, it does offer a means of operationalizing the undoubted connections between individual human behavior and broader environmental and social contexts (Duncan et al., 1996).

We saw earlier how it is possible to analyze case-control data in a spatial setting, where the "cases" are individuals with some disease or disability and the "controls" are a group of individuals with the same broad characteristics (age, gender, social class, for example) but without the disease or disabling condition. We used a spatial analytic method to see if we could detect "clusters" of cases, relative to controls. Case-control methodology is widely used in quantitative health research, where data for individuals are used to examine the influence of one or more "risk factors," adjusting or "controlling" for the influence of other variables. We shall see in later chapters how these ideas are used in particular research contexts. But here we introduce a standard statistical method, the *logistic regression* model and look briefly at an example.

Logistic regression is appropriate where the variable to be explained takes on one of two values. This contrasts with conventional regression analysis, where the dependent variable is continuous, and with the Poisson regression model, where it comprises a set of counts. Logistic regression is therefore used where we wish to look at factors and variables that predict a binary dependent

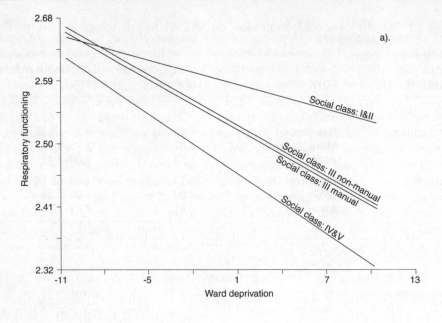

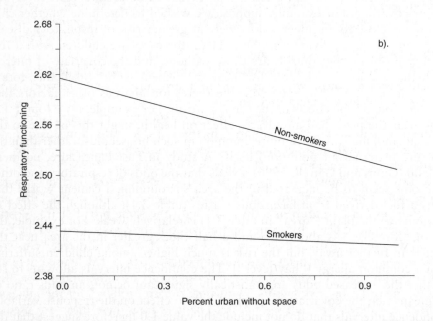

Figure 3.11 Multi-level modeling of respiratory functioning: a) effects of deprivation and social class and b) urban index and smoking (Source: Jones, K. and Duncan, C. (1995) Individuals and their ecologies: analysing the geography of chronic illness within a multilevel modeling framework, *Health and Place*, 1, 27–40, reproduced with kind permission of Elsevier Science)

Table 3.1   Results of logistic regression analysis of health survey in part of East Lancashire

Symptom	Covariate	Odds ratio	Confidence interval (95%)
Sore throat	Area	1.83	1.15–2.91
Blocked nose	Hay fever	4.81	2.23–10.36
	Mold in house	2.82	1.25–6.35
	Area	1.58	1.00–2.48
Sore eyes	Hay fever	26.76	11.26–63.60
	Mold	2.04	1.05–3.98
	Area	2.67	1.01–7.09

'Area' takes on two values, according to whether or not a child lives near the cement works
Source: Ginns and Gatrell (1996: 633)

variable (here, case or control). The regression coefficients that emerge from a logistic regression analysis may be converted into *odds ratios* that reveal the "odds" of disease or illness, given the particular value of an explanatory variable or covariate. For example, suppose we wished to determine whether children living close to a cement works were at greater risk of respiratory disease than a similar group living elsewhere. Here, the cases are children with a respiratory problem, the controls are those without, and the covariate of interest is whether or not the children live near the cement works (itself therefore a binary variable). We can enter into the regression analysis other factors that may "confound" the effects of the cement works. For example, if we found that living near the plant had an adverse effect on health, might this be due to the fact that smoking was much more common in such households, rather than the exposure (external air pollution) itself? A study in East Lancashire, northern England (Ginns and Gatrell, 1996) reveals that the odds of respiratory and other symptoms seem to be increased in the area surrounding a cement works that burns a fuel derived from hazardous wastes (table 3.1), although the effect of proximity to the plant ("area" in table 3.1) is relatively weak. For example, the risk of "sore eyes" is about two and a half times greater in the area near the plant than further away, but the risk is much higher among children suffering from a common allergy ("hay fever"). The confidence intervals allow us to see whether the increased odds are statistically significant or not; an odds ratio of 1.0 means that the covariate in question has no effect on the response variable. Confidence intervals that do not include the value 1.0 therefore suggest that the covariate is significant at a specified level of probability.

Geographical Information Systems and Health

A good deal of the spatial analytical work outlined in the above sections has been paralleled by the development of geographical information systems (GIS).

Such systems first emerged in the 1960s, but it is only in the past 20 years that the science and technology have taken off, and only relatively recently that health geographers have begun to exploit these developments.

A GIS is a computer-based system for collecting, editing, integrating, visualizing, and analyzing spatially referenced data. Such data comprises two forms: the locational element and associated "attributes." For example, we might have data on the residential addresses (locations) of people suffering from stomach cancer, of different ages and sex (attributes). We might want to link or integrate such data with other databases, such as those concerned with environmental quality. Exploring or modeling the associations between the health and environmental data is complicated, because the data sets will have been collected for different sets of locations. As another example, we might have data on the locations of health centers (which will have attributes such as number and type of staff employed there, and opening times), along with data on patients treated there (who will also have locations and attributes, such as health status); we might wish to model the flows of people to health centers.

Let us consider some examples to illustrate what is possible. Consider first the incidence of malaria in Africa. In the South African province of KwaZulu-Natal the disease has had devastating effects. Malaria control teams equipped with global positioning systems (GPS) receivers have obtained the latitude and longitude of nearly 35,000 family homesteads in parts of the province, and they store "green cards" under the roof eaves to record the application of insecticides as well as results of blood smears for parasite detection; all these data now form part of a geographically based malaria information system (Sharp and le Sueur, 1996). The system also identifies the locations of clinics to treat affected people, and the GIS allows the user to place *buffer zones* around these clinics in order to determine which homesteads are relatively remote from these sources of health care (box 3.3). Maps of malaria prevalence may also be produced, permitting control agencies to direct resources more systematically to areas of need. This approach is now becoming widespread in the developing world, such as in the Tigray region of Ethiopia, while the use of other satellite-based systems (remote sensing) is beginning to be used in disease-monitoring (Thomson et al., 1996). For Africa as a whole, Snow and his colleagues (1999) have sought to produce good estimates of the numbers of people killed each year by malaria. They derive a climate suitability model for the disease, based on climate data for weather stations that are then interpolated onto a 5×5 km grid. Those areas are deemed most suitable for transmission of malaria in which temperature is between 18°C and 22°C (64–70°F) for at least five months, and where rainfall is on average 80 mm (3 ins) per month also for at least five months. Data on population distribution, by age, are obtained for over 4,000 administrative areas and also interpolated onto a grid. Last, making some assumptions about age-specific mortality rates, and applying these to small area population estimates in regions of differential malaria risk (as assessed by climate), Snow and his team suggest that there were likely to have been just under one million deaths from malaria in the continent as a whole during a single year (1995). Others have derived similar estimates, but a GIS "provides a more rational basis for defining disease burden, which is transparent in its inputs, assumptions, limita-

BOX 3.3   *Collection and manipulation of digital spatial data*

Most paper maps have a system of "geo-referencing" or spatial coordinates that allow us to determine a grid reference, or perhaps a latitude and longitude, for a particular feature. We can also think of linear features (such as roads and rivers) as comprising sequences of coordinates, while areal units comprise ordered sequences of point coordinates that form a boundary. In order to transfer such paper-based information into electronic form we need a means of "capturing" this coordinate information. This is done by a process known as "digitizing," whereby a hand-held cursor on a special table (in which is buried a fine grid of copper wires) tracks the features of interest and transmits the coordinate data to a computer file. Clearly, this can be time-consuming! Fortunately, we do not always have to collect our own digital spatial data, since national mapping agencies do this themselves. In Britain, the Ordnance Survey (OS) makes available digital data on a variety of spatial scales, including large-scale data on the locations of individual properties ("ADDRESS-POINT") and centre-lines of roads ("OSCAR").

In the USA the Bureau of the Census provides comprehensive national spatial data coverage. Topologically integrated geographic encoding and referencing (TIGER) files give grid references for street intersections, as well as "topological" data that indicate which blocks lie on which side of each street segment.

In situations where such databases do not exist, or where paper-based maps are inadequate, we have to rely on other forms of data capture. At broad regional scales we may find remote sensing – the use of satellites – of use, but more locally we use global positioning systems (GPS), as in the South Africa malaria study described in the text.

It is frequently of interest to construct "buffer zones" around point features (such as factories or hospitals). From an epidemiological perspective this allows us to examine disease incidence, not on the basis of fixed areal units but within a given radius of a site. From the viewpoint of health care planning it gives us a simple means of assessing access to health care facilities. We can also construct buffer zones of fixed radius around linear features such as roads or power lines. This too could be used to monitor accessibility but a more common use is to assess disease risk along such features; is the incidence of asthma elevated along busy main roads, for example?

A difficulty with buffer zones is that the size of the zone is usually rather arbitrary, and the results depend on how they are defined. This is another manifestation of the modifiable areal unit problem; see pp. 53–4 above.

See Martin (1996) and Longley et al. (1999) for further details.

tions and caveats" (Snow et al., 1999: 635). It needs to be borne in mind that the difficulties in collecting data (whether on health or demography) in parts of the developed world are immense.

The possible health effects of living near high-voltage power lines, a source of electromagnetic radiation, has generated considerable controversy in recent years (see, for example, Day et al., 1999). The issue is difficult to resolve, both because of the problem of measuring the exposure and because the associated health effects (thought to be leukaemia, brain, and other cancers) are rather rare. Some Finnish work (Valjus et al., 1995) has used GIS to assess possible exposure. Finland has a computerized population and buildings register; the locational coordinates of all properties are known, as are residential histories from the 1960s onwards. The authors digitized maps of high voltage power lines (box 3.3) and linked these to the buildings and residences files. It was thus possible to calculate the number of people living in buildings located within particular distances or buffer zones of the digitized power lines. Given the locations of the lines, along with average annual currents and distances to the center points of buildings it is possible to estimate the exposure of individuals, over time, to what is known as magnetic flux density. Results (figure 3.12) show the distribution of the Finnish population living within 150 metres of power lines in 1989, and their level of exposure. About 15,600 persons were exposed to a field strength greater than the normal background level, and 530 to a significantly greater level. These are small fractions of the total population (about five million persons) and, given the rarity of the possible health events linked to such exposure it would be difficult to draw conclusions about cause (exposure to electromagnetic radiation) and effect (disease). Nonetheless, this is a good example of the value of GIS in a health-related context, made possible only because of the detailed population and buildings database, common throughout Scandinavian countries.

A third example is drawn from a study of the levels of lead in the blood of children in the city of Syracuse, in up-state New York (Griffith et al., 1998). Here, the authors are concerned to describe and analyze the spatial distribution of such levels; is there an association with transport routes or with residential quality (the age or value of housing, for example)? Lead finds its way into the bloodstream via emissions from leaded gasoline (petrol), lead-based paint and drinking water (historically transmitted through lead pipes), among other routes. A normal background level is about $0.5\,\mu g/dl$, and health experts call for intervention if levels exceed $10\,\mu g/dl$; interest therefore centers on levels that exceed this. Using a GIS it is possible to match addresses to streets and census units (tracts and blocks); the dot map of highly elevated levels (figure 3.13) shows two areas of concentration, each running from north-west to south-east. The authors place buffer zones around the cases with elevated levels and it is clear from both visual and statistical analysis that there is no association between high levels and proximity to roads.

There is clear evidence of spatial autocorrelation in the blood-lead levels when these are aggregated to either census tracts or census blocks (smaller units); high levels in one census unit tend to be associated with high levels in nearby areas. The authors fit a regression model to the data, in an attempt to

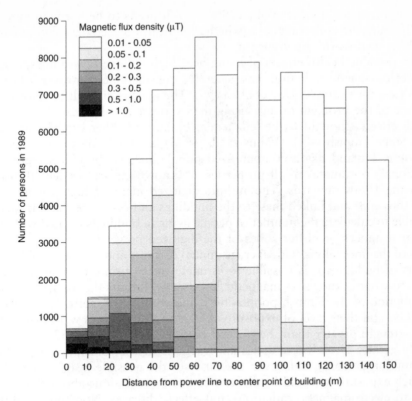

Figure 3.12 Distribution of people in Finland ever having lived within 150 meters of power lines, and their estimated average annual personal magnetic flux densities (Source: Valjus, J., Hongisto, M., Verkasalo, P., Jarvinen, P., Heikkila, K., and Koskenvuo, M. (1995) Residential exposure to magnetic fields generated by 100–400 kV power lines in Finland, *Bioelectromagnetics*, 16, 365–76, reproduced with kind permission of Wiley-Liss Inc, a subsidiary of John Wiley & Sons, Inc.)

find covariates that predict blood-lead levels. Although the findings depend upon the scale – a common feature of geographical analysis – there are striking associations with mean house value (figure 3.14) and this proves to be a good predictor. The conclusion is that levels are elevated in the poorer neighborhoods and that intervention, such as improving the quality of the housing stock, and ensuring it is maintained, needs to be focused here.

Two final examples of the use of GIS relate to a health care planning context, one in part of North Carolina, USA (Walsh et al., 1997), the other in Canada (Scott et al., 1998). Walsh and his colleagues construct a database comprising the locations of hospitals, the digitized road transport network, and the locations of patients (by zip code). Data on the transport network includes the length of links from one location or "node" to another, so that distances between zipcode areas and hospitals can be computed readily; while making assumptions about speed of travel also allows travel time estimates between hospitals and places of residence to be computed. Data are available on the number of beds

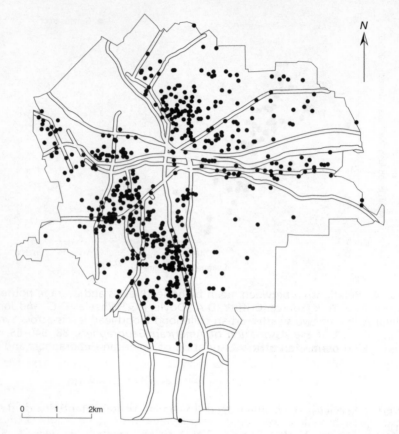

Figure 3.13 Children with high blood-lead levels in Syracuse, New York, with "buffer zones" (150 ft) around roads (Source: Griffith, D.A., Doyle, P.G., Wheeler, D.C., and Johnson, D.L. (1998) A tale of two swaths: urban childhood blood-lead levels across Syracuse, New York, *Annals of the Association of American Geographers*, 88, 640–65, reproduced with kind permission of the Association of American Geographers and Blackwell Publishers Ltd)

available in each hospital, and a network algorithm within the GIS works outwards from each hospital, assigning to those hospitals segments of the road network (and associated populations attached to such links) until the bed supply is exhausted or all patients have been allocated to hospitals. In this way it is possible to design a set of health service areas that minimizes aggregate travel time. The value of a GIS approach here is in simulating possible changes to the health care delivery system. For example, what happens if a hospital closes, or if demand in a sub-region grows? In the North Carolina example the authors allow patient population to double; this shrinks the supply areas, leaving large rural areas unserved. While this is not necessarily a realistic scenario it does suggest the *decision support* capabilities of the GIS. Moreover, one can investigate the delivery of particular health services for subsets of the population, substituting population over 75 years, say, as a measure of demand

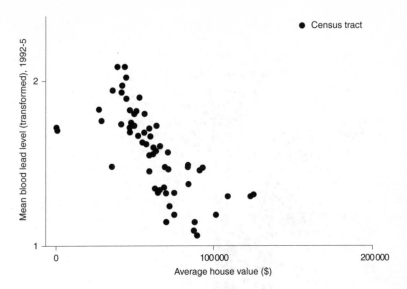

Figure 3.14 Relationship between mean blood-lead levels and average house value in Syracuse, New York (Source: Griffith, D.A., Doyle, P.G., Wheeler, D.C., and Johnson, D.L. (1998) A tale of two swaths: urban childhood blood-lead levels across Syracuse, New York, *Annals of the Association of American Geographers*, 88, 640–65, reproduced with kind permission of the Association of American Geographers and Blackwell Publishers Ltd)

for geriatric services, or women aged 15–45 as those potentially demanding obstetric care.

A further example comes from work on access to hospitals in Canada that offer specialist facilities for the treatment of patients suffering an acute stroke (Scott et al., 1998). It is recommended that patients suffering such an episode must be evaluated and treated within three hours of the stroke onset; assuming that the immediate hospital care requires one hour then two hours remain for transport from place of onset to hospital. The Canadian research suggested that this corresponds approximately to a distance of 105 kilometers (65 miles) – an assumption that could be modified using more detailed digitized road networks – and so this radius can be superimposed on the locations of all 110 hospitals offering the requisite care. Using population data from the census it is possible to identify the proportion of the population living within 105 kilometres of such centres. Results suggest that 85 percent of the total population is so served, though this access varies with socio-economic status. Those who are better educated are closer than those who had completed less than nine years of education. Similarly, while 70 percent of families with incomes in excess of CAN$50,000 were within the critical distance, only 54 percent of those with incomes less than CAN$20,000 had such access. We need, therefore, to disaggregate population "need" to look at the needs of particular social groups. Similar work has been conducted in the United States (Love and Lindquist, 1995).

Interpreting the Geography of Health: Qualitative Approaches

Until quite recently, quantitative studies of disease incidence tended to focus almost exclusively on aggregate spatial patterns; a good example was that considered in Chapter 1, where the spatial distribution of AIDS in Uganda was of interest. More recently, as we have seen, attention has switched to quantitative analyses that are based on individual events, and the spatial arrangement of point patterns that represent collections of individuals with disease have been studied. The danger, however, is that the analytical methods give little attention to what the points or dots on the map really represent. The dots are not inanimate objects; they are real people, and while the ways in which they are arranged spatially may shed some light on disease causation the approaches we have explored so far give no consideration at all to the feelings, experiences, beliefs, and attitudes that the individuals have. We need to have these people speak to us rather than reducing them to a collection of dots on a map.

There are several ways in which we can engage with these individuals in a *qualitative* way. We might want to ask them a set of questions, interviewing them at length but allowing them to shape the conversation rather than dictating it ourselves. We might want to spend long periods of time with them, participating in their daily lives in order to better understand their experiences of illness or disability. How we do this, and how we analyze the material we collect, is the focus of this section. But while there was more emphasis on data analysis in the section on quantitative methods, more weight is given here to methods of data collection.

We have already noted that there is no simple one-to-one relationship between theoretical stance and methodology; qualitative methods might be used by social interactionists, structurationists, postmodernists, and others. They are unlikely to find much use among positivists, however, since the latter set store by measurability and objectivity. Those using qualitative methods do so because they seek an understanding of human beliefs, values, and actions, and they do so from the standpoint of an equal partnership between the researcher and the researched. As we see below, there is a considerable variety of methods for collecting qualitative data.

Interviews

Structured questionnaires seek to elicit information from a set of individuals and yield information that is amenable to quantitative analysis, perhaps permitting some generalizations to be made from the sample to a wider population. In a health context they invariably ask for information about age and sex, as well as asking about health status. But they do not give the respondent any opportunity to really "respond," presenting instead a series of questions that invite only a limited number of responses. It might be possible to make room on the questionnaire for the respondents' own comments (and this is frequently done) but an alternative strategy would be to engage the respondent in a conversation, the purpose of which would be to cover a number of themes but not simply to run through a list of fixed questions.

This conversation, or *in-depth interview* (sometimes referred to simply as a depth interview), gives the interviewees more scope to describe and explain their own experiences of health and illness, perhaps raising issues that had not occurred to the interviewer. The aim is not to obtain an "objective" account of health status, nor does it treat the interviewees as "objects." Rather, the approach is avowedly subjective, treating those being interviewed as people whose values, beliefs, and feelings are to be respected and valued as legitimate sources of data to inform the study. The essentially hierarchical nature of the questionnaire schedule, where the researcher is "in charge," controlling the process in a position of some power, thus gives way (ideally) to a more equal relationship in which the conversation is shaped as much by the interviewee as by the interviewer. This form of relationship is given particular importance by those engaged in feminist research.

Those being interviewed are chosen as an indicative rather than a representative sample. For example, if we wished to explore the uses people make of health services, and of particular interest to us were urban and rural settings, and gender variations in usage, we might wish to interview a small number of men and women living in towns and in small villages. An interview schedule would be prepared, perhaps outlining a brief list of key themes to be explored during the conversation. We need to emphasize that the sample is not a "probability" one, in the sense that there is a known probability of an individual being sampled; rather, the aim is to select individuals who may provide a rich body of information concerning the problem being tackled.

A good example of this style of research was the study conducted by Cornwell (1988), while work by Eyles and Donovan (1986) is also illuminating. They present an account of how people living in the anonymized West Midlands town of "Mossley Green" make sense of illness and opportunities for health care. They draw upon in-depth interviews with 30 individuals as a way of uncovering what it means to be ill and to need health care in this particular locality. A number of themes emerge from their field work, including: how health and illness are perceived; how lay people think about the "causes" of sickness; and the social context of health and illness. In discussing perceptions of health and illness the notion of "biographical disruption" referred to in Dyck's study (Chapter 1) comes to the fore, as when one interviewee tells us that "In the morning, I'm in agony, trying to get my breath. Half me life I've suffered from fibrositis of the spine. . . . I've got pins in me ankles and thumb and in cold weather, they're terrible" (Eyles and Donovan, 1986: 418). Perceptions of health care are shaped by social structure and expectations, as when reductions in public expenditure are thought to be a problem: "We've two lovely hospitals in the town . . . they have to send you out of the beds because they haven't got enough staff" (Eyles and Donovan, 1986: 420). Possible causes of sickness (such as smoking and diet) are thought to be both under the control of the individual but also shaped by industrial history: "I have to walk through the industrial site and there is a white car that has been parked there and it has gone all yellow from the pollution. . . . there's a foundry and you can taste it sometimes, very sweet tasting" (Eyles and Donovan, 1986: 423). Clearly, while some of the narratives carry few clues as to the spatial setting, others are couched within a very place-specific context.

The methodology employed by Eyles and Donovan is therefore explicitly qualitative, in the sense of "learning first-hand about people, observing and analyzing 'real life' and studying actions and learning about ideas as they occur" (Eyles and Donovan, 1986: 416). They seek to take the "common-sense" ideas of their sample, to appraise and analyze this material, and to derive constructs and categories from the observational data. This is what others have referred to as *grounded theory* (literally, theory grounded in, or derived from, the data); see below, page 86.

Sabatier (1996) has conducted interviews with migrant women from southern Africa, examining their vulnerability to HIV infection together with the mechanisms they adopt for coping with the infection. Some interviews reveal the need for sexual bartering in order to supplement low incomes, but as one Zimbabwean woman explained: "Why are we women called 'prostitutes' when we come to town looking for work? They say we are responsible for the spread of AIDS, but what about these men? How else am I going to feed my children, which the father fails to support?" Another interviewee, a 20-year-old market trader in a Tanzanian town, reported on the harassment she had received from a market inspector and the support she enlisted in producing video evidence of his harassment: "They showed him the video and told him they will show it in the community if he doesn't leave us alone. They tried to educate him about AIDS. Condoms aren't the only way to fight AIDS!" The picture Sabatier paints via these women's narratives is one of strength and resistance. Women take part in many informal social networks, within which they are not simply powerless or victims; rather, they are creative agents who make effective use of the often scarce resources available to them.

As we saw in an earlier section, one approach to geographical epidemiology is a "scientific" study that collects data on disease incidence using official sources, then maps and analyses these for evidence of clustering or elevated disease risk. In contrast, there is a growing literature on popular or "lay" epidemiology, one that gives priority to the accounts, perceptions, and fears of those who worry that there is a raised incidence of disease, possibly caused by environmental contamination. We consider here one example (see also box 8.2 below, page 230), since it makes extensive use of qualitative methods. This study is of a small Australian coastal community, Oceanpoint, in New South Wales, comprising fewer than 3,000 people. Whittaker (1998) reports results from 88 in-depth interviews, covering a wide range of people of different ages and social groups. These people (representing a relatively large sample for this kind of study) were identified in a "snowball" manner, by making contact with community organizations, members of which assisted in the recruitment of other informants as well as by simply walking round the community requesting interviews from people who also suggested others to talk to. In 53 of the 88 taped interviews comments were made about health and environment, of which 23 referred specifically to cancer. The popular view is that cancer incidence in Oceanpoint may be linked to leaching of contaminated water from a former site for the disposal of hazardous waste, or perhaps the waste ("tailings") from a former sandmining operation. As one mother of a 31-year-old man who had died of leukaemia suggested, "two doors up from here was another young man who had cancer and the street behind us another young man. I think the eldest would have been thirty-five . . . They all died with cancer within twelve months and to me that's a terrific coincidence." She suggested that the creek where her son played was contaminated and that this might be implicated, though her account also acknowledged that this was not the sole explanation; for example, "the young man that died two doors up had come from in the country some-where and he hadn't grown up in Oceanpoint" (quotations from Whittaker, 1998: 318–19). Underlying this lay epidemiology are feelings about lack of control and an inability to prevent powerful industrial interests from taking decisions that affect the community. As one 39-year-old woman expressed it: "They always tried to dump everything in the Oceanpoint area I think because we've been out of the way and you know, we haven't always had a louder voice to kind of combat things" (Whittaker, 1998: 322).

This example reveals something of the value of detailed interviews, in which such qualitative methods are seen as the only way of revealing local community understandings of health problems. If we want to get to grips with people's health worries we can only do this by engaging with them at length. Having said this, there are obvious dangers in the way respondents are sought and in verifying the information they reveal, and few public health researchers would want to leave unchallenged some of the "evidence" presented. They would argue that different cancers have different possible causes, that we know little about the sources or duration of exposure to environmental contamination, and that unless we place a very local study in a broader regional context we have no way of knowing whether the popular concerns are supported by "hard" facts. This tension between different forms of *evidence* is a recurring theme in contemporary health geographies.

As a final example consider Davidson's (2000) work on women with agoraphobia. Agoraphobia is an abnormal fear of open spaces, or of being in public places. As Davidson (2000: 33) explains, "experiencing a reduction in the locations in which they can move comfortably (or at all) sufferers may eventually be unable to leave their homes without experiencing an incapacitating level of anxiety." Davidson interviews in depth ten agoraphobic women living in Scotland. She finds that the home is for some a safe location, though for others it is so "secure" that it may become imprisoning. She contrasts this with the shopping trip (the "malady of the mall") where the architecture and potential for social interaction generates anxiety. For some, it is not simply space itself that threatens; rather, the feeling of being surrounded by others. As one of her respondents explains, "I'm fine in an open space. As long as there's no' a whole lot o' people there I'm fine" (quoted in Davidson, 2000: 32).

There are methodological problems in conducting interviews, and the interview setting is invariably power-laden, in the sense that the interviewee may feel in an inferior position relative to that of the interviewer. This is less the case than with a questionnaire (and certainly less so in feminist research), but there is still a risk of academic "voyeurism" in which the researcher gains valuable data and the respondent benefits not a jot. The interview is a social relationship involving both the researcher and the researched. As such it is often difficult and uncomfortable, for both participants, but as Parr (1998: 348) observes, we can hardly expect an in-depth interview not to reveal in-depth feelings. As a result, the interview might be a very emotional experience for both parties. The taping of interviews can be extremely problematic. Apart from the obvious risks of mechanical failure, the instrument can itself be a barrier to interaction or, for the mentally ill in particular (the focus of Parr's own empirical work), perceived as yet another means of surveillance. Moreover, the spatial setting of the interview cannot be ignored; in what place should it occur? The home setting may be comfortable for the interviewee, though issues of the safety of the interviewer must be addressed.

Participant Observation

Some qualitative researchers argue that it is only possible to collect rich data by actively participating in, and observing, daily human activity. This style of collecting data, known as *participant observation*, grew out of early-twentieth-century anthropology and ethnography, where some researchers immersed themselves in the lives of the subjects of study, in as unobtrusive a way as possible. In a health context such a method has been used to study hospitals, wards within hospitals, or groups of individuals such as alcoholics. At one extreme the researcher becomes a "total" participant (being employed as a nurse perhaps) and thereby adopting a covert research role; at another extreme, the researcher separates him or herself from the group being researched, so that while deeply immersed in their daily lives s/he maintains social and emotional separation.

There are obviously limits to the numbers of people whose worlds can be observed; between five and eight people is usual, 15 at most. Issues of temporal sampling are also important; is it better to observe over limited periods, or for longer, continuous periods? When collecting data, whether overtly or covertly, the researcher "acts largely as a camera, scanning and recording detail wherever she/he happens to be focusing, while also recording sounds and spoken language from a broader range" (Grbich, 1999: 134). Given the need to be unobtrusive, the researcher should write up notes "off-site" and as soon as possible after leaving the site of study. The data collected might also include photographs and video recordings, but again care must be taken not to have the equipment intrude into the setting.

Let us examine briefly two pieces of empirical work that have included distinctively ethnographic approaches to health, from a geographical perspective. Both, as it happens, are concerned with AIDS. First, Michael Brown has looked at the local politics of AIDS in Vancouver, Canada, becoming a participant observer at the Pacific AIDS Resource Centre and at AIDS Vancouver (an AIDS service organization). In addition, he also explored the oral histories of 120 people involved in local AIDS politics. These included "elected city officials, provincial and city bureaucrats, a host of volunteers, family members, and people themselves who are living with AIDS, tracking the locations of their political engagement with the AIDS crisis" (Brown, 1997: 16). Three broad questions were asked: "What is it you do or have done in the local response to AIDS?" "How did you come to do these things?" "How have you seen things change or have they stayed the same?" (Brown, 1997: 28). To illustrate Brown's work, consider one group of people he talked to at length – "buddies". A buddy is someone who volunteers practical help and emotional support to a person living with AIDS or HIV. As Brown (1997: 123) puts it, "buddying is a form of citizenship defined through locations where elements of family, home, and state relations are combined in places across the city." It emerged as a response to the AIDS crisis, and in particular the failure of both family and state to deal adequately with the needs of those infected with the virus.

Where, then, does "buddying" take place? Brown shows that locations comprise a mix of public and private spaces, including simply meeting for coffee or dinner, going for a walk or drive (perhaps to the beach or park), or meeting in either the client's or the buddy's home. As one told him:

> "We walk miles! Just miles and miles and miles. I have walked from Spanish Banks over the Burrard Bridge back to Stanley Park. And we went to Lighthouse Park. . . . And oh, man, did we climb! We found a nice, quiet little cove and we talked. . . . He is one of these people who has absolutely no family support. He has only told one other person that he tested positive. And he no longer sees that person after telling them – after knowing them 20 years. Personally I think that's bloody shameful" (cited in Brown, 1997: 139–41).

For many of the clients, the value of a buddy is "just being there;" but Brown shows that this "does not just mean being *any*where" (Brown, 1997: 154).

Paul Farmer's research on AIDS in Haiti is another fine example of ethnography or interpretive anthropology, but set within a political economy perspective. He attempts to relate large-scale (both historical and contemporary) events to the lived experience of ordinary village people and their families and neighbors. His methodology is, principally, that of participant observation, and his research has involved spending several years among people living in one village, Do Kay, a community of about 1,000 people that had originally comprised refugees displaced in the 1950s by a hydroelectric dam. The experiences of three people living with (and dying of) AIDS are examined in depth. One, for example, a man called Dieudonné, has a structuralist explanation of the illness: "What I see is that poor people catch it more easily. They say the rich get [AIDS]; I don't see that. But what I do see is that one poor person sends it on another poor person" (Farmer, 1992: 12). Quoting others, he suggests that "the poor, in their popular wisdom, in fact 'know' much more about poverty than does any economist. Or rather, they know in another way, in much greater depth" (Farmer, 1992: 262–3). But the misfortunes of ordinary "lay" people are linked to those of the communities they live in as well as the appalling economic position of Haiti itself.

Haiti's place within a West Atlantic economic system whose hub is the United States has played a major role in shaping the epidemiology of HIV and AIDS in the country. These economic linkages have included the tourist industry, and other qualitative work has demonstrated that impoverished "beach boys" have been infected by, and in turn infect, those with whom they have sexual contact. Economically driven male prostitution, involving a largely North American clientele, has played a pivotal role in the spread of HIV to Haiti (rather than, as some have claimed, being spread from a Haitian "source" to North America). Indeed, Farmer (1992: 261) goes so far as to suggest: "Given that unequal relations between the Caribbean and North America have contributed to the current epidemiology of HIV, an analogous exercise leads to a somewhat analogous observation: the map of HIV in the New World reflects to an important degree the geography of U.S. neocolonialism." HIV has diffused "along the fault lines of economic structures long in the making" (Farmer, 1992: 9). These deeper structures cannot be observed but are nonetheless real. Farmer's qualitatively based methodology is, therefore, couched firmly within a structuralist framework.

Focus Groups

A focus group is a collection of a small number of people, typically between four and twelve, that meets to discuss a topic of mutual interest, with assistance from a facilitator or moderator. Usually, the group members are "key informants", that is, they represent particular positions or interests. The discussions are informal and consist essentially of exchanges of views and opinions and the swapping of personal experiences. The atmosphere is intended to be non-intimidating and to permit discussion that does not necessarily result in agreement or consensus. As with a series of interviews, much thought must be given to the location and physical setting in order to create the optimal environment for interaction. The discussions are taped and transcribed (as with in-depth interviews, possibly a lengthy and complex process) and analyzed. Assembling an appropriate mix of people is not always straightforward; nonetheless, focus groups do serve to give people a sense of "ownership" in any research, and if findings are fed back to them it prevents any sense in which they feel they are used merely as "data fodder".

A good illustration of the method comes from Garvin's (1995) report on the development of a partnership between female academics and community researchers in order to research women's health and health needs. The setting is Dickenson County, in the Appalachian region of south-western Virginia; this is a rural area, with a history of coal-mining but now suffering from high unemployment. The cooperative research project sought to "give a voice to Appalachian women's experiences, draw attention to their concerns and quality of life, reduce excess burdens of illness, facilitate the development of a culturally sensitive research process, and create a planning process that could be diffused to surrounding counties as women learned from and about other women" (Garvin, 1995: 274).

Six focus groups, comprising in total 44 women, were organized according to location or area of interest. Discussion was focused on a number of issues, including: women's health concerns; health-seeking behavior; perceived barriers to accessing health care; and perceived quality of local health services. Following transcription, the academic researchers identified and summarized key themes that were then sent to the community researchers (including nurses, social workers, and health visitors) for comment. Themes were grouped into three sets of factors affecting use of services. These were: predisposing factors or individual attitudes (such as religious beliefs); enabling factors, concerned with local or federal policies that permit or constrain health behaviors (transport provision, for example); and reinforcing factors that reward or modify subsequent behavior (such as the quality of service). In general, users felt that services were inaccessible, inconvenient, and poorly delivered. They were well aware of the need to adopt a "healthy" lifestyle, but argued that the lack of facilities for exercise, the lack of choice in local food stores, and the expense of getting screening and other preventive care all acted as barriers from acting on knowledge. This is a theme to which we return in the next two chapters, but meanwhile this study serves as a good example of how to elicit information on health services from the people they are supposed to serve.

Longhurst (1996) too has made use of focus groups, as means of exploring the experiences of pregnant women in Hamilton, New Zealand. Only two women took part in some of the focus groups, which were designed to examine what sorts of places such women spent time in, both before and during pregnancy. There were some findings of geographical interest; for example, that "[r]epresentations of women as 'hysterical' and physically incapacitated serves to construct and inscribe them as belonging within the confines of the domestic realm, as unsuitable for the rigours of public life" (Longhurst, 1996: 146). However, "place" does not emerge as a very strong theme.

Qualitative Data Analysis

We have looked at three broad classes of method for collecting qualitative data. How is such material to be analyzed? In contrast to quantitative approaches, in which the analysis of data follows on, and is quite separate, from the collection of data, qualitative data analysis is often bound up closely with data collection. Typically, data collection is used to refine, discard, and then reformulate ideas, quite differently from the formal testing of hypotheses used in much quantitative work.

Following, in part, Grbich (1999) I shall discuss methods of analysis very briefly under three headings. These are: *enumerative*, where documentary material is analyzed, typically using content analysis; second, *investigative* methods, such as discourse analysis; and third, *iterative* methods, where interview material is analyzed. Clearly, the "distance" between the researcher and the subject decreases from enumerative through iterative methods.

Content analysis examines documents such as text or visual materials, specifying elements or units (such as words, phrases, or images) to be analyzed and noting the relative frequency of particular codes. Computer packages may be used to retrieve units and compile frequency tables, although for some this is a rather "atomistic" approach that breaks continuous content into discrete lumps. Discourse analysis sees texts as historically produced materials that embody and reflect power relations. A "discourse" is a theme or subject made up of "discursive practices;" for example, the medical records of an individual, and how he or she interacts with a health professional are discursive practices within a wider discourse of, say, patient health. A key issue is to trace the power relations that have shaped the discourse; such relations might be between doctors and nurses, or those professionals engaged in health promotion and the public they seek to influence. Discourse analysis thus draws to some extent on Foucault's post-structuralism, nicely illustrated in an analysis of health service reform in the UK in the early 1990s (Moon and Brown, 1998).

I think it is worth devoting a little more attention to iterative analysis, since the analysis of interview transcripts is increasingly common in health geography. With iterative approaches the researcher collects data via interviews or observation, transcribes this information (for example, typing the interview from the taped conversation), and reflects on what it has to say about the topic under investigation. This is a preliminary analysis that annotates transcripts, highlighting in the text some ideas in a process known as "open coding." It requires a detailed, slow, and careful reading of the text, with the analyst's comments in the margins or sections of text highlighted carefully. (There is a variety of computer packages now available to manage large volumes of transcript and other qualitative material, but it must be stressed that these cannot do any genuine analysis; this is a task for the researcher.) Emerging concepts and themes then shape further data collection, and data collection halts when the researcher sees no obvious gaps. The skill in qualitative data analysis lies, then, in the coding of the text and the subsequent identification of themes or concepts. Broad themes may need disaggregating, in a process known as "axial coding." But there is a continual interaction between the analyst and the data, in order to ensure that nothing is missed and that the codes and themes "fit" the data. This process is clearly inductive, in the sense that one moves from the data to an emerging set of concepts in order to develop a theory. Since this is very much grounded in the data, the inductive nature of the process is known as a *grounded theory* approach (Grbich, 1999: 171–80).

Recently, MacKian (2000) has introduced some novel forms of visualization into the qualitative analysis of health narratives. In essence, this has led to her devising "maps" as heuristic devices to allow us to understand a respondent's experiences and what has shaped these. Drawing on her research into people with chronic fatigue syndrome (myalgic encephalomyelitis or ME) she demonstrates very convincingly that these visual representations reveal the structure of experience and the relative locations of the factors that relate to this experience. MacKian's respondents in turn confirmed that the maps were valid representations of their worlds.

Concluding Remarks

The distinction between quantitative and qualitative methods is long-established, and it cannot be denied that those researching the geography of health have tended to use one or the other set of tools, a conclusion mirrored in health research in general. Others, however, take a more pragmatic and eclectic stance, using whatever methods are appropriate to the problem under investigation. Indeed, for some researchers, the mixing of both quantitative and qualitative methods has proved singularly useful, with insights from in-depth interviews adding color and explanatory power to quantitative studies. This mixing of methods (and, perhaps, sets of data) is sometimes called *triangulation*. Different methods may reveal a consistent picture, though in some cases the results of a qualitative analysis may run counter to those from a statistical one. Triangulation may be *simultaneous*, where quantitative and qualitative data collection and analysis run in parallel, or *sequential*, where the collection and analysis of qualitative material follows, or precedes, a quantitative approach.

Within health research in general, and medical research in particular, until quite recently quantitative data analysis held sway, mirroring the dominance of a positivist approach to research. However, even in biomedical research the value of qualitative research has come to be recognized. Within the geography of health the same has happened, and the use of qualitative methods has possibly superceded that of quantitative methods. For many geographers, however, myself included, both sets of methods can shed light on most research questions. We shall see evidence of this eclecticism in the rest of this book.

This chapter ends the section of the book that deals with approaches to, and methods for, conducting enquiries into the geography of health. We turn next to a series of substantive issues. These deal first with the social environment, where it is quite clear that the different styles of enquiry compete for attention. In the second substantive section we look further at issues linking the physical environment to health outcomes; here too the different methods of explanation raise their heads, though perhaps not as prominently as earlier in the book.

FURTHER READING

For quantitative studies of the geography of health you should begin with Cliff and Haggett (1988), a classic text that sets out a variety of cartographic and spatial analytic approaches, including many that I have not considered. It is full of examples, most of which relate to the developed world and many to Iceland.

Modern approaches to spatial data analysis, including those for handling both point and area data, are considered in Bailey and Gatrell (1995); this gives several health-related examples. It includes a software package and some sample data sets.

On GIS, a good general introduction is Martin (1996). I have co-edited a collection of relevant papers: see Gatrell and Löytönen (1998), but for another overview see also

Gatrell and Senior (1999). The edited collection containing this chapter has much of generic value.

An excellent introduction to empirical Bayes estimation is in Langford (1994), while other examples are in Elliott et al. (1992b). Other applications of these ideas in geographical epidemiology include research on sudden infant death syndrome (SIDS) in North Carolina (Getis and Ord, 1992) as well as further analysis of the child mortality data for Auckland (Marshall, 1991; Bailey and Gatrell, 1995).

An overview of some spatial analytic methods based on point process analysis is given in Gatrell et al. (1996), though Bailey and Gatrell (1995) have further details and Diggle (1993) is a primary source. All of these contain references to additional techniques and applications. On multilevel modeling see Jones (1991) for a pedagogic overview, and Jones and Duncan (1995) for a clear exposition and good example.

The book edited by Flowerdew and Martin (1997) has several chapters on research methods and techniques. In particular, the chapters by Valentine on interviewing, Cook on participant observation, and Crang on the analysis of qualitative material, are all valuable sources for the qualitative researcher, and merit a close reading.

There are numerous texts emerging on the use of qualitative methods in health research. I have found Grbich (1999) of particular value. I have not had the space to look in detail here at the ethnographic studies by Brown (1997) and Farmer (1992), but both deserve a careful reading. For a good example of applying qualitative methods (in-depth interviews) to local concerns about landfill-siting decisions, work that draws upon Beck's "risk society" (Chapter 2 above), see Wakefield and Elliott (2000). Helpful comments on sampling in qualitative health research are given in Curtis et al. (2000). See also the collection of papers on the geography of health and impairment in *Area* (2000, 32, 7–78).

PART II

HEALTH AND THE SOCIAL ENVIRONMENT

CHAPTER 4

INEQUALITIES IN HEALTH OUTCOMES

In this chapter I first describe some of the patterns of inequality relating to health outcomes, particularly as they manifest themselves in the developed world. There is, of course, a large number of possible health "outcomes" on which I could draw, including mortality (both all-cause and numerous specific causes), life expectancy, and many indicators of morbidity or disability. Inevitably, therefore, the account will be selective, but I hope that the guide to further reading (page 133) will suggest possible sources for evidence not covered here. There is a huge literature on health inequalities, particularly from those interested in the sociology of health and illness. Much of this focuses on inequalities that are expressed in terms of social class, gender, and ethnicity. I shall touch on all three, but wish to give particular emphasis to the spatial patterning of health inequalities and to explanations that give some prominence to place.

I adopt an approach that is based on geographical scale, beginning with some patterns of variation at the international scale, before examining some of the broad regional differences in selected countries. I then consider some work on quite local, small-scale variations in health outcomes. Although it is difficult to draw a firm distinction between descriptive and explanatory accounts (if for no other reason than that most studies try to explain their descriptive findings!), I then turn in the second half of the chapter to reviewing some of the competing explanations for health inequalities. As we shall see, some authors consider these to lie in varying health behaviors, while others believe that health inequalities can only be explained by structural factors that are deeply embedded in human societies. Others offer different explanations for these inequalities.

I think it is also worth pointing out that while the term "health inequalities" has become accepted, we really mean "health inequities." The distinction is important. Inevitably, there will be unevenness or variation in health outcome, whether from place to place or along some other dimension. But what really matters, and what "inequity" implies, is the fact that these differentials may be avoidable, and should be capable of being narrowed; their existence is, in a sense, unethical. Fundamental issues of social justice should be borne in mind in a reading of this and the following chapter.

Patterns of Inequality

Health Inequalities: International Comparisons

I want to begin by drawing attention to broad contrasts between countries in the developed and developing worlds. There are numerous health indicators that might be used to describe these contrasts (Curtis and Taket, 1996). Mortality data are an obvious source, though the variable quality of death certification makes this problematic. Morbidity data are even more poorly reported in some parts of the world (Phillips, 1990). Some comparative work uses data on life expectancy at birth, or on Disability Adjusted Life Years (DALYs) though since both are based on mortality data they need to be treated with some caution. DALYs are defined as the sum of years of life that are lost due to premature mortality and years lived with a disabling condition (adjusted for severity); thus they take account of morbidity and disability as well as mortality.

Research conducted at the Harvard School of Public Health in association with the World Health Organization (Murray and Lopez, 1996), known as the Global Burden of Disease Study, has used DALYs in order to quantify and compare ill-health across the world. Results (Murray and Lopez, 1997) show that the developed world ("established market economies," or EMEs) accounts for only 7.2 percent of the global burden of disease, while 90 percent of this burden is carried by the developing world (figure 4.1). This is because countries in sub-Saharan Africa and large parts of Asia (India in particular) have high incidence of, and mortality from, infectious and parasitic diseases, respiratory infections, nutritional deficiencies, and maternal and perinatal disorders. In contrast, in the EMEs, cancer, cardiovascular disease and mental illness (especially severe depression) account for the burden of ill-health.

While these statistics offer a reasonably up-to-date snapshot, patterns of mortality and morbidity do change over time; in particular, the contribution made by different causes of death is highly variable from country to country, and over time, a feature that has been encapsulated in the concept of *epidemiological transition* (sometimes known more generally as "health transition"). In essence, this proposes that at an early stage of development causes of death such as parasitic, infectious, and nutritional diseases account for the burden of mortality, but then these give way to non-communicable diseases. Diseases that account for high rates of infant mortality give way to the degenerative diseases, such as heart disease and cancer, in adulthood and old age (Phillips, 1990; 1991). As overall levels of fertility and mortality fall, diseases that are common in the developed world begin to account for a greater proportion of deaths in developing countries.

This is a very simplified and brief account, but we can illustrate the epidemiological transition with reference to countries in south-east Asia (Phillips, 1990; 1991; 1994). In Hong Kong infectious and parasitic diseases (mostly diphtheria, enteric fever, and dysentery) accounted for nearly a quarter of deaths in 1951, a proportion which had fallen to 3 percent in 1988 (table 4.1).

Disability Adjusted Life Years (DALYs) by region
(% total DALYs worldwide, 1990)

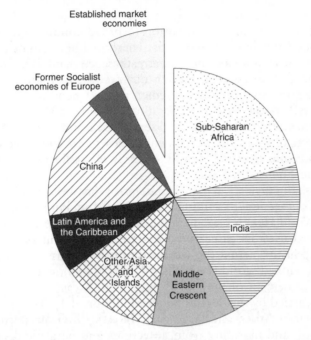

Figure 4.1 The global burden of disease: Disability-Adjusted Life Years, 1990 (Source: Murray, C.J.L. and Lopez, A.D. (1997) Global mortality, disability, and the contribution of risk factors: Global Burden of Disease study, the *Lancet*, 349, 1436–42, reproduced with kind permission of The Lancet Ltd)

Table 4.1 Epidemiological transition in Hong Kong

Year	*Infectious and respiratory*	*Cancer*	*Cardiovascular*	*Parasitic diseases*
1951	23.6	4.2	5.5	27.4
1961	16.2	12.1	7.4	14.8
1975	4.0	24.2	26.6	15.8
1988	3.0	29.8	28.6	17.2

Percentage of all deaths

Source: Phillips (1990: 45)

The pattern of cancer deaths was the reverse; these now account for 30 percent of deaths, compared with 4 percent in 1951. Rates of heart disease and stroke have also increased. This is evidence of very rapid epidemiological transition. Thailand shows a similar pattern; the crude death rate from tuberculosis dropped from 47 to 12 per 100,000 between 1970 and 1983, while the crude death rate for all cancers rose from 13 to 27 per 100,000. Malaysia and Singapore offer a similar picture (Phillips, 1991).

The rate of epidemiological change differs greatly from country to country; some, such as Thailand and Hong Kong, show an "accelerated" transition, while others, mostly in Africa, show a "delayed" transition; infectious disease remains the major burden. Moreover, we should not assume that the transition is homogeneous within a single country; there are major regional contrasts, such as urban-rural differences in health outcomes, within both developed and developing countries. In addition, the emergence of new infectious diseases (most obviously AIDS), the continuing high incidence of malaria and other infectious disease, and the re-emergence of diseases such as tuberculosis means that we cannot point (even in the developed world) to an end stage of the epidemiological transition in which simply diseases of "affluence" remain (Smallman-Raynor and Phillips, 1999).

As we shall see later, particular attention has been paid, within the health inequalities literature, to varying incidence of circulatory diseases: those involving blood circulation (box 4.1). Data from different developed countries reveal huge variations from country to country, suggesting that an homogeneous late stage in the epidemiological transition is a gross over-simplification. For

---

BOX 4.1   *Circulatory diseases*

There is not space in this book to discuss in any detail the nature of specific diseases, but since diseases involving blood circulation have been the focus of considerable attention in the health inequalities field they merit brief consideration here.

The most common form of circulatory disease arises from the deposition of fatty material (atheroma) in the linings of the arteries that supply blood to the heart muscle (myocardium). The resulting narrowing of the arteries and insufficient blood supply is known as *ischaemic heart disease*. In an extreme case, the narrowing of the arteries causes a *myocardial infarction* or coronary thrombosis (commonly known as a "heart attack"). Diets that are high in saturated fats have a high ratio of cholesterol to protein and increase the volume of "low density lipoproteins" in the blood. Since these low density lipoproteins transport fats to muscles they are a risk factor in heart disease. Tobacco products also serve to degenerate coronary arteries and are therefore an additional risk factor. The generic name for diseases of the heart and blood vessels is *cardiovascular disease*.

Narrowing of the arteries in the brain causes "stroke," a form of *cerebrovascular disease*. Since the brain cannot function without supplies of oxygen, supplies that are provided by the blood flow, strokes are frequently associated with symptoms of paralysis, usually on one side of the body or face. Stroke shares some of the same risk factors as ischaemic heart disease, including diets rich in saturated fats, and smoking.

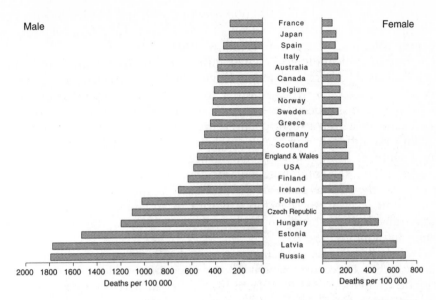

Figure 4.2 Age-adjusted mortality from circulatory diseases (various dates between 1991 and 1995)

example, the death rate from circulatory diseases among men aged 55–64 years in the USA is substantially higher than in Canada and it is nearly twice as high in Germany than in neighbouring France (figure 4.2). Although rates among women are generally one third of those for men, the same broad international variations remain.

One Europe or Many?

As implied in the previous section, there is heterogeneity of health outcomes, even within the developed countries of Europe. It is evident from figure 4.2 that there is a very clear gap between eastern Europe and the rest of the continent. In this next section we consider some of the further evidence relating to cross-national variations in Europe. In particular, I want here to draw particular attention to evidence of an "East-West" divide in health outcomes. Since 1970 mortality rates in eastern Europe have increased while those in all countries of western Europe have declined (Watson, 1995). This is gender-biased, however, in that while both men and women in eastern Europe do worse than in the west the differential is far greater among men (figure 4.3: a) shows the graph for men, and b) that for women). Hungary had, in 1990, the highest death rates among both men and women, but mortality rates among eastern European women had actually improved slightly over the 20 year period. The increased death rates among men were due to a number of causes, including heart disease, respiratory disease, and cancer. But these data refer to the period 1970–90, and pre-date the break-up of the former Soviet Union; what of more recent trends?

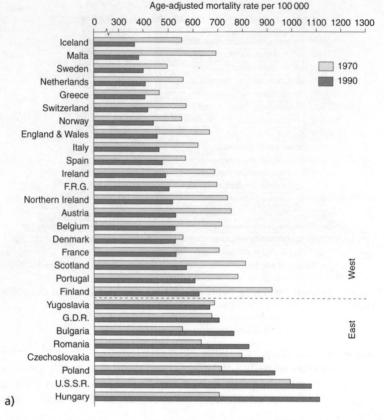

Figure 4.3 Mortality rates in Europe for a) men and b) women aged 25–64 years: 1970 and 1990 (Source: Watson, P. (1995) Explaining rising mortality among men in Eastern Europe, *Social Science and Medicine*, 41, 923–34, reproduced with kind permission of Elsevier Science)

Life expectancy at birth, for men in the European Union as a whole, rose from 72 years in 1985 to about 73.5 years in 1994. Life expectancy in the former USSR had always been lower than in the west; and in the Russian Federation the figure was 63 years in 1985. However, life expectancy has declined sharply since then (figure 4.4a), and in 1994 was under 60 years of age; although not quite so dramatic, the same is true for women (figure 4.4b). (See page 99 for figure 4.4). This figure for Russia as a whole masks considerable spatial variation, however, as analysis of data for 52 regions reveals (Walberg et al., 1998). The greatest declines in life expectancy among men were in the north of the country, in more urbanized areas. Much of the decline is due to high death rates from cardiovascular disease, but attention needs to be given to wider structural changes, as we see later.

There are, then, striking differences between countries, but we need to ask whether those living within such countries are at equal risk from early death?

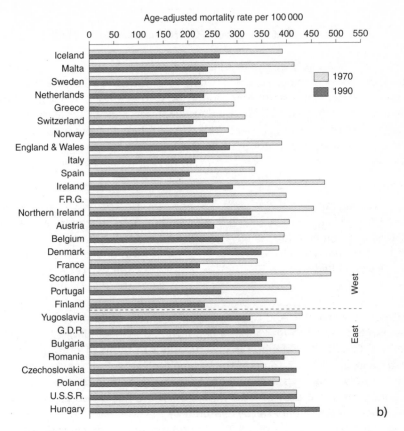

Figure 4.3 *Continued*

To what extent are there regional differences according to socio-economic status (box 4.2)?

We may point up some international contrasts between those people working in manual occupations, compared with those in non-manual classes (Kunst, 1997). From figure 4.5 we see variations in the probability of a middle-aged man dying between ages 45 and 65, for both the USA and selected European countries. For those in non-manual occupations there is relatively little difference, while for those in manual occupations the probabilities of mortality in eastern Europe (again, for Hungary in particular) are up to twice as high as in western countries. This pattern is repeated across other dimensions of social status, such as level of education. Kunst (1997: 149) further shows that in 1990 the ratio of mortality rates among men aged 30–44 in manual, compared with non-manual, occupations, was 2.25 in the Czech Republic and 2.89 in Hungary. Comparable figures for the USA and England and Wales were 1.42

BOX 4.2   *Socio-economic status*

In the USA inequality of health outcome is commonly assessed with reference to either race or income. Income is widely used, since the US census collects such data on a decennial basis. In other countries, such as Britain, no income data are generally available. As a result, another dimension, that of social class or socio-economic position, is used instead. This is usually based on occupation. In Britain a six-fold classification is adopted, as the following table (with example occupations) illustrates:

Social class		Example occupation
I	Professional	Doctor, lawyer, accountant
II	Managerial	Sales manager, teacher, nurse
IIIN	Skilled non-manual	Clerk, secretary
IIIM	Skilled manual	Bus driver, chef, carpenter
IV	Partly skilled	Farm worker, security guard
V	Unskilled	Labourer, cleaner

Often, classes I and II, and IV and V, are combined, while there is a simple distinction between non-manual (I, II and IIIN) and manual (IIIM, IV and V) occupations. Clearly, this scheme ranks occupations in terms of prestige and social status and, implicitly, in terms of the income each attracts. It is not without problems, however. Married women who do not work outside the home are usually assigned the social class of their husbands. Those out of work, or children, or the elderly, are also imperfectly captured by this scheme.

As an alternative, we might use household tenure or educational level as measures of status, or car ownership as a possible index of "wealth." There are problems with all of these indicators, but each in its way is strongly associated with mortality and morbidity.

and 1.46, respectively. These differences are maintained for all leading causes of death.

At city level recent work has sought to make comparisons between the major (mostly capital) cities of some European Union states. The Mégapoles Project has derived a set of over 40 indicators of health status or health determinants. These include mortality data, road traffic accidents, tuberculosis, HIV/AIDS, low birth weight, and teenage conceptions. Some examples are shown in table 4.2 (page 100). Inevitably, different recording systems mean that comparisons are difficult. Nonetheless, we can point to some broad differences between the "north" and "south." Southern European cities such as Athens and Lisbon show high rates of infant mortality and deaths from road traffic accidents, while some northern cities (for example, Amsterdam) tend to have higher incidence of TB and AIDS.

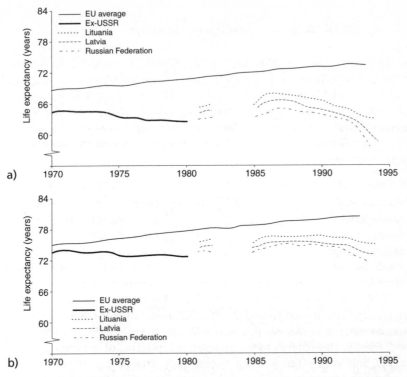

Figure 4.4 Life expectancy at birth for a) men and b) women in European Union and selected countries of the former USSR (Source: "Health Inequalities: Decennial Supplement", ed. F. Drever and M. Whitehead, 1997, Office for National Statistics, Crown Copyright 2000, reproduced with kind permission of the Office for National Statistics)

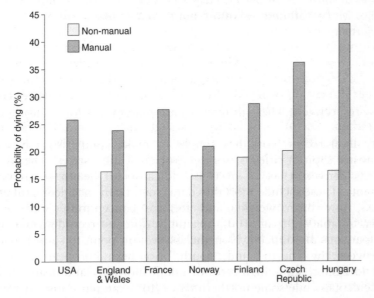

Figure 4.5 The probability of a man dying between the ages of 45 and 65 (Source: Kunst (1997) *Cross-National Comparisons of Socio-economic Differences in Mortality* (Erasmus University, the Netherlands), reproduced with kind permission of the author)

Table 4.2 Health indicators for major cities in some European Union countries

City	SMR[a]	IM[b]	RTAs[c]	TB[d]	AIDS[e]	LBW[f]	TCR[g]
Amsterdam	28	6.0	40	32	3,080	5.1	n.d.
Athens	n.d	8.3	133	n.d	88	7.4	n.d.
Berlin	28	5.5	35	20	978	6.5	n.d.
Brussels	30	6.2	47	n.d	1,764	6.3	n.d.
Copenhagen	45	5.7	87	7	n.d.	5.1.	n.d.
Dublin	23	4.6	65	10	n.d.	5.7	21
Helsinki	26	3.2	27	14	5	n.d.	17
Lazio (Rome)	22	6.3	102	12	1,034	5.9	12
Lisbon	30	6.2	161	49	754	n.d.	n.d.
London	24	6.1	39	31	1,042	8.2	52
Lyon	20	4.1	70	10	669	6.6	13
Madrid	20	4.2	77	30	2,465	6.1	8
Oslo	25	4.8	42	19	n.d.	5.9	23
Stockholm	20	4.3	41	9	578	4.2	26
Vienna	30	5.4	34	28	561	6.3	n.d.

[a] Standardised mortality rates under age 65 (deaths per 10,000)
[b] Infant mortality rates (deaths under 1 year, per 1,000 live births)
[c] Road accident deaths per million residents
[d] Incidence of tuberculosis per 1,000 residents
[e] Cumulative cases per million residents
[f] Proportion of births under 2,500 g.
[g] Estimated teenage conceptions per 1,000 population aged 15–19 years
The letters n.d. indicate no data

Source: http://www.elcha.co.uk/holp/

Health Inequalities: Regional and Class Divides

I want to turn now to some of the evidence for health inequalities within particular countries. As in the previous section, some consideration will be given to social class and gender differences, though I wish here to introduce ethnicity as a further dimension of health divides. I concentrate on evidence relating to England and Wales, and the USA.

It is, I think, useful to begin by observing that, despite the fact that the literature on health inequalities has mushroomed over the past 20 years, health inequalities have been documented since at least the early part of the nineteenth century (Macintyre, 1998). Edwin Chadwick's report into the working conditions of working-class people in Britain showed how life expectancy varied by both social class but also by area (table 4.3). The social class gradient is striking, but note too the higher life expectancy in the more affluent southern areas of Rutland and Bath; life expectancy there was higher for all social classes.

Much more recently, broad regional differences in health outcome in England have been revealed among regional health authorities (now transformed into a smaller set of regional offices of the NHS). Data on life expectancy at birth for

Table 4.3 Mean age at death of members of families belonging to various social classes in England (1840s)

District	Gentry/professional	Farmers/tradesmen	Labourers/artisans
Bath	55	37	25
Rutland	52	41	38
Bethnal Green	45	26	16
Leeds	44	27	19
Manchester	38	20	17
Liverpool	35	22	15

Source: Macintyre (1998: 21)

males in 1992 (figure 4.6) show that life expectancy in southern regions was on average three years higher than in the north (Charlton, 1996). Although on average life expectancy for women was five years more than for men, the same broad regional differences are observed for women too. At a finer geographical scale, Raleigh and Kiri (1997) have mapped life expectancy for both men and women for the set of English health authorities, using data for 1992–4. Again, there is considerable variation, with the "healthiest" districts (such as Cambridge) reporting a life expectancy of 76.6 years (81.1 for women) and the least healthy (Manchester) a life expectancy of 69.9 (76.7) years. Put simply, a male born in Cambridge could expect to live six and a half years longer than one born in Manchester, while a female could expect to live about four and a half years longer.

A classification of local authorities according to socio-economic characteristics indicates that the differences in life expectancy are magnified, varying from 75.8 (80.4 years for females) in the most prosperous areas to 71.7 (77.5) in port areas. A similar health "divide" exists if we look at how morbidity varies among these broad social areas (table 4.4, page 103). Rates of heart disease and mental illness are much higher in traditional industrial areas than in the most prosperous ones.

Broad regional contrasts in health outcomes were noted in the so-called "Black Report" (DHSS, 1980), which has over the last 20 years stimulated a considerable body of research into health inequalities in Britain and Europe. Among many important studies the work of Shaw and her colleagues (1999) deserves careful reading, since it looks in a very imaginative way at premature mortality (deaths under 65 years) for parliamentary constituencies in Britain. In particular, it contrasts those constituencies with a population of one million persons aged under 65 years, which have the highest SMRs, with a similar set having the lowest SMRs. There are many findings of interest here; to cite just one, the Scottish constituency of Glasgow Shettleston has an SMR of 234 (1991–5 data), while Buckingham in southern England has an SMR of just 71. Further, this gap has widened since the 1980s (Shaw et al., 1999: 118).

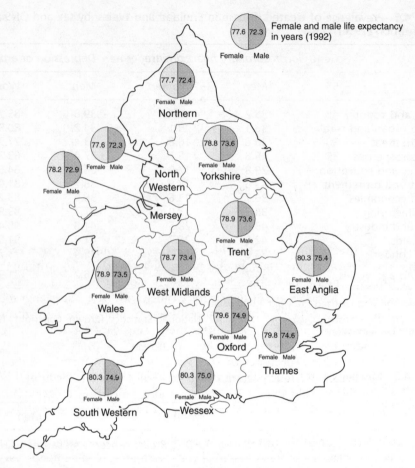

Figure 4.6   Life expectancy for men and women in England and Wales, 1992

Some of this geographical work highlights broad areal contrasts; we need to begin to understand the extent to which such contrasts simply mirror the composition of the population. Do certain regions have low life expectancy, or high mortality, simply because poorer people, or those belonging to a different ethnic group, live there? These are difficult questions, on which the whole chapter touches, but we can begin to answer them by examining some of the evidence on social class and other gradients.

Mortality rates in England and Wales show clear evidence of a social class gradient, a gradient which is observed for all causes and which is consistent across most of the major causes of death. For example, mortality rates among women and men in lower social classes are between 50 and 70 percent higher than in professional and managerial social classes (table 4.5), while for heart disease the gradient for women is even more marked; the mortality ratio is 2.7. Most cancers show higher mortality in lower status groups, though note that breast cancer shows no evidence of a gradient (and skin cancer mortality – not

Table 4.4 Prevalence of treated disease in England and Wales, by sex and ONS area classification (1996)

	Coronary heart disease		Depression or anxiety	
	Men	Women	Men	Women
Coast and country	35.2	19.4	39.8	85.7
Mixed urban and rural	33.1	20.1	34.2	80.9
Growth areas	30.6	16.8	33.4	77.7
Most prosperous	25.8	15.7	27.0	63.2
Services and education	29.8	17.9	26.6	64.5
Resort and retirement	33.6	18.8	38.7	81.0
Mixed economies	36.7	23.6	36.6	83.2
Manufacturing	35.3	23.0	37.7	83.9
Ports and industry	43.0	28.5	41.8	95.4
Coalfields	42.0	26.7	41.2	91.6
Inner London	24.8	14.7	35.4	73.7
England and Wales	34.7	20.8	36.2	81.9

Data are age-standardized rates per 1,000 patients

Source: Health Statistics Quarterly, Office for National Statistics (Crown Copyright 1999), reprinted by permission

Table 4.5 Mortality rates (ages 35–64 years) by social class in England and Wales (1986–92)

	Women			Men	
	All causes	CHD[a]	Breast cancer	All causes	CHD[a]
Professional and managerial	270	29	52	455	160
Skilled non-manual	305	39	49	484	162
Skilled manual	356	59	46	624	231
Partly skilled and unskilled	418	78	54	764	266
Ratio of partly skilled and unskilled to professional and managerial	1.55	2.69	1.04	1.68	1.66

[a] Coronary heart disease

Source: Independent Inquiry into Inequalities in Health (1998: 13)

shown in the table – also tends to be more common among professional people; see Chapter 9).

In Britain, a series of studies of those working for the government in office-based work ("civil servants") has revealed clear social class gradients in health outcomes. These studies (known as the Whitehall studies) began in 1967 and have tracked more than 17,000 individuals. Early work (Whitehall I) focused on mortality and revealed that the lower the grade of staff, the higher is the standardized mortality for heart disease (and, indeed, for virtually all diseases); there is a four-fold difference in rates between the highest and lowest grades of occupation (Marmot et al., 1978). In other words, even within a single occupational group of non-manual workers there are major differences in mortality risk. Later work (Whitehall II) has concentrated on morbidity and has confirmed the social class gradient for heart disease, bronchitis, and self-reported health status (Marmot et al., 1991).

In the United States there is a large variety of health and health-related social indicators that can be used to paint a contemporary picture of health divides (Miringoff and Miringoff, 1999). Let us consider evidence on inter-state differences in three such indicators: life expectancy, low birth weight, and births to teenagers. Nation-wide, life expectancy at birth for white males is 73.9 years, but only 66.1 years for black males, a gross disparity. Geographically, life expectancy is higher in the western and mid-western states of Utah, Colorado, Minnesota, and Wisconsin (where some counties show average life expectancy of over 76.5 years). Contrast this with parts of the rural south or inner city areas (Bronx, New York, or Baltimore, for example), where men can expect to live 12 years less. The poorest life expectancy is in six counties of South Dakota (61 years), reflecting the presence of native North Americans on reservations. Also of interest is to contrast neighboring Washington D.C. (62.2 years) and Fairfax, Virginia (76.7 years).

There are also considerable variations in the proportion of infants who are of low birth weight (under 2,500 grams). At least 8.7 percent of all births in states in the south (Mississippi, Louisiana, Alabama, South and North Carolina, and Tennessee) are of low birth weight, compared with under 6 percent in the north-west (Washington and Oregon) and in New Hampshire (4.8 percent). The same contrasts are observed for births to women aged 15–19 years. Rates (per 1,000 such women) are over 70 in states such as Mississippi, Arkansas, and the south-western states of Arizona, New Mexico and Texas, but under 33 per 1,000 in the New England states of Massachusetts, Maine, Vermont, and New Hampshire.

As above, we need to know whether these geographical differences reflect social composition; in particular, in the USA to what extent are these broad differences attributable to the varying life chances of black and white people? There is a clear patterning of health outcome according to "race" or ethnicity (the American literature tends to adopt the former term, though the latter is more common elsewhere). For black American adults aged 25–44, the all-cause mortality rate among both men and women is over twice that of white adults (Sorlie et al., 1995). These differences are spatially variable, however. For example, the SMR among black men aged 15–64 living in Harlem, New York City is 411 (1989–91 data), but only 181 among black men in the southern state of Alabama (Geronimus et al., 1996); for black women the same figures are 338 and 189 respectively. Given that the "standard" is 100 for white people then mortality rates are over four times as high for some black populations. Put differently, a 15-year-old black person in Harlem has a 37 percent chance of surviving to the age of 65; this compares with a 77 percent chance for blacks as a whole, and 87 percent for whites. More recent evidence (Geronimus et al., 1999) confirms this and indicates that for both black women and men there is a rural "advantage" to living in southern states, compared with cities such as Chicago, New York, and Detroit where both homicide and AIDS have led to worsening mortality between 1980 and 1990 (figure 4.7: a) shows the graph relating to men; b) that relating to women). Clearly, a focus on race/ethnicity in the absence of "place" will not suffice.

Do these stark contrasts remain when we adjust for income? In other words, do black people whose family income is similar to that of whites have similar health outcomes, or do the differences remain when allowance is made for these influences? The answer seems to be that they do. Regardless of race, those with lower incomes have consistently higher mortality than those who have higher incomes; this is true for both men and women (table 4.6). Yet, across any particular category, blacks have higher mortality rates. Using data from the US Longitudinal Study, black male mortality rates (among 25–44 year olds) are 36 percent higher than for whites, and 69 percent higher among black women, even when adjustment has been made for employment status, income, and education (Sorlie et al., 1995).

Similar inequalities according to ethnicity emerge in Britain, though particular care needs to be taken with "non-whites" in this context since it is problematic to assign an "ethnic group" to an individual, while health outcomes vary substantially from one group of non-whites to another. Allowing individuals to categorize their ethnicity according to family country of origin, and looking at data on self-reported health status, evidence suggests (figure 4.8) that non-white ethnic minorities as a whole are 20 percent more likely than whites to report poor or only fair health. This masks important differences, however, and even the crude "south Asian" category cannot distinguish the significantly poorer health of Pakistanis and Bangladeshis from the much better health of "Indians." The latter is itself a very crude grouping, masking the very different cultural, religious, and geographical origins of people; for example, Muslims tend to report worse health than Hindus or Sikhs (Nazroo, 1998).

Table 4.6 Mortality rates in USA (25–64-year-olds) in 1986[a]

Annual income ($)	Women		Men	
	White	Black	White	Black
<9 000	6.5	7.6	16.0	19.5
9 000–14 999	3.4	4.5	10.2	10.8
15 000–18 999	3.3	3.7	5.7	9.8
19 000–24 999	3.0	2.8	4.6	4.7
>25 000	1.6	2.3	2.4	3.6

[a] Age-adjusted death rates, per 1,000

Source: Macintyre (1998: 23)

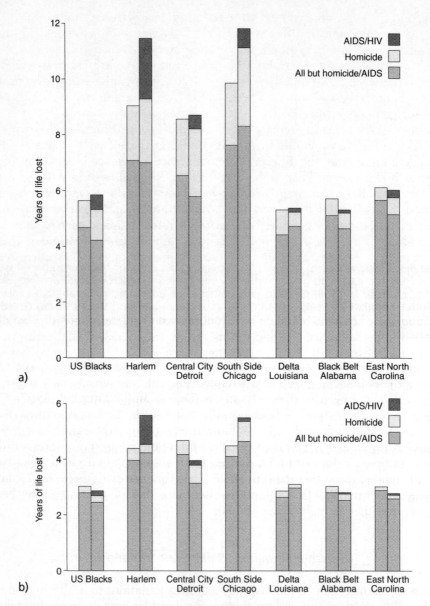

Figure 4.7　Years of life lost in areas of USA in 1980 and 1990, by cause: a) black men; b) black women (Source: Geronimus, A.T., Bound, J., and Waidmann, T.A. (1999) Poverty, time, and place: variation in excess mortality across selected US populations, 1980–1990, *Journal of Epidemiology and Community Health*, 53, 325–34, reproduced with kind permission of the BMJ Publishing Group)

Another key theme within the health inequalities literature has been to assess whether such inequalities have improved, or worsened, over recent years. In general, the evidence suggests the latter. For example, in Sweden mortality among men employed in professional occupations has declined markedly since the early 1960s; on the other hand, mortality among those employed in manu-

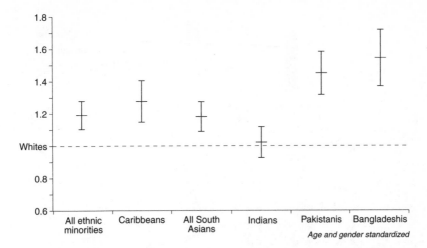

Figure 4.8 Relative risk of non-whites reporting fair or poor health compared with whites: risk for each sub-group, with 95 percent confidence interval around each (Source: Nazroo, J. (1998) Genetic, cultural or socio-economic vulnerability? Explaining ethnic inequalities in health, in Bartley, M., Blane, D., and Davey Smith, G. (eds.) *The Sociology of Health Inequalities*, reproduced with kind permission of Blackwell Publishers)

facturing and industry worsened through the 1970s, and although the gap between the two social groups has declined since, it was substantially wider in 1990 than it had been in the early 1960s (Drever and Whitehead, 1997: 52). As another example, while infant mortality rates in the US have declined since 1950, for both black and white infants, the ratios of these rates have not improved dramatically. The ratio of black/white neonatal mortality rates (deaths within the first 28 days of life) has risen from about 1.5 in 1950 to over two in 1991, while the postneonatal mortality rate (infants dying between 28 days and one year) among black infants is also twice that of whites (having been nearly three times as high in the 1960s).

Health Inequalities: Small-Scale Variations

Some research has been conducted on small-area variations in health outcomes, looking at both mortality and morbidity. We consider here some studies taken from parts of South Africa, USA, Spain, New Zealand, and Britain.

In South Africa, Donald (1998) has looked at tuberculosis notifications in two adjacent, disadvantaged communities in Cape Town. Donald uses a GIS to map the cases within 39 "Enumeration Districts," where incidence varies from 78 per 100,000 to 3,150 per 100,000. There is considerable small-scale spatial variation, and further analysis shows a clear association with mean household income (figure 4.9).

In northern Manhattan (covering Central Harlem and Washington Heights) Durkin and her colleagues (1994) have examined small-area variations in rates of severe injury (that is, resulting in hospitalization or death), to children under

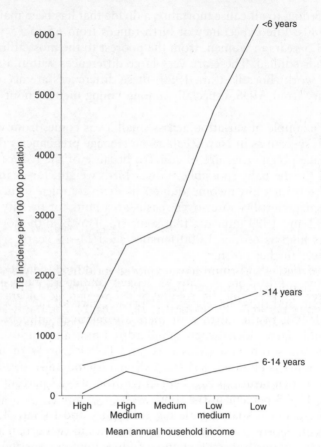

Figure 4.9 Relationship between incidence of tuberculosis and income in parts of Cape Town, South Africa (Source: Donald, P.R. (1998) The epidemiology of tuberculosis in South Africa, *Novartis Foundation Symposium: Genetics and Tuberculosis*, ed. D. Chadwick, and G. Cardew, John Wiley & Sons Ltd, reproduced with kind permission of the Novartis Foundation)

17 years, between 1983 and 1991. Data were analyzed by census tract, and were classified according to cause. Across the study area as a whole, unintentional injuries had the highest incidence (57.8 per 10,000 per year), followed by falls (16.4) and motor vehicle injuries (12.6). But this masks considerable spatial variation; for example, the incidence of all injuries varies from 3.9 to 160.2 per 10,000, while for burns it varies from zero in some small areas to 276.4. We cannot ascribe the high rates to small numbers of children in those tracts (see the discussion on p. 53), because tracts with fewer than 100 children were excluded from the analysis. What we can say is that, on average, nearly three out of a hundred children each year suffer or die from major burns in one small area.

Research in Barcelona, Spain (Borrell and Arias, 1995: see Drever and Whitehead, 1997) shows a wide gap between the wealthiest and the poorest districts according to all-cause mortality, a divide that has been maintained since the early 1980s. Life expectancy at birth ranges from 65 to 75 years in men, and 75 to 81.5 years in women, from the poorest to the most affluent of the 38 neighborhoods studied. These are very large differences within a single city. In more recent work (Borrell et al., 1999) these differences are accounted for by high mortality from AIDS, especially among young men without much formal education.

A further example of variation across small areas comes from work done by public health specialists in New Zealand on teenage pregnancies (Wilson et al., 1996). Teenage pregnancies are of concern because of the risk of adverse outcomes, both for the baby (possibly a lower birth weight, for example) and the mother (possibly on a low income and less likely to complete formal education). A map of teenage fertility rates over census area units for the city of Auckland between 1992 and 1994 indicates that some small areas in the south of the city have rates as high as 65 per 1,000 females aged 11–19 years, while rates are generally lower further north.

Two sets of studies in Britain have explored local health divides. Research in Glasgow, Scotland, has compared the health experiences of those living in relatively affluent, and those in relatively deprived, parts of the city (Sooman and Macintyre, 1995). In the north-west there are relatively affluent communities (West End and Garscadden) where standardized mortality ratios are quite low (83–127). In the south-west (Mosspark and Pollok), areas of poorer public housing, mortality is elevated (SMRs between 123 and 154). In addition, those reporting their own health as excellent range from 18–28 percent in the north-west to only 11–23 percent in the south-west.

On a more detailed spatial scale my colleagues and I (Gatrell et al., 2000) have contrasted deprived and affluent communities in two urban areas, Salford and Lancaster, in north-west England. A questionnaire survey of over 770 respondents allowed us to paint a fine-grained picture of perceived health, and to map individual responses using postcoded addresses. Such maps show considerable local variation in health outcomes. The task then is to interpret and explain such variation, at whatever scale is considered, and we turn attention now to this problem.

Explaining Inequalities in Health Outcomes

Inevitably, having outlined some of the huge range of descriptive findings relating to health inequalities it has not been easy to avoid hinting at some of the possible explanations. We have seen, for example, that there are socio-economic gradients, and that health outcomes vary by sex and ethnicity. We need now to consider some of the types of explanation that have been put forward to shed light on such inequalities.

We can picture health status in terms of a series of layers of influence, with a set of fixed factors (age, heredity, and sex) at the core, surrounded by a series of factors that might be modifiable. The first layer comprises lifestyle factors representing behavior(s) that may or may not be conducive to good health. The second represents social networks and community influences, while the third comprises living and working conditions, as well as access to services and facilities in the local area. Finally, there is a set of wider structural determinants representing the influence of macro-economic and broad-scale social conditions on individual health (Whitehead, 1995).

This schema is an advance on the traditional distinction between behavioral (lifestyle) and material conditions; in other words, a simple "human agency" and "structure" dichotomy. Yet it too is an oversimplification, since the different layers impact on each other, as when our individual "choices" about health behaviors (whether or not to smoke, for example) are in part affected by wider-scale structural influences (such as the freedom of cigarette manufacturers to advertise). Interestingly, from a geographical perspective there is something of a correspondence between these layers and spatial scale, since we move from individual influences, to local or neighborhood effects, through to broader national scale determinants. With these observations in mind, we now examine some of the evidence relating to each "layer."

The Programming Hypothesis and the "Life Course"

As we shall see in the next section, the conventional causal model in much chronic disease epidemiology emphasizes adult "risk factors" or behaviors, such as smoking, diet, and exercise, as determinants of health status. This has been challenged by a number of workers and research groups, who claim that diseases such as diabetes, chronic bronchitis, and cardiovascular disease are biologically "programmed" in early life (in the womb or in infancy). In particular, the work of David Barker and his colleagues has produced a considerable body of evidence in support of this theory. Barker has collected together some of this literature (Barker, 1992; 1994) and we consider it briefly, restricting attention to cardiovascular disease.

From a geographical point of view, there is compelling evidence in support of Barker's hypothesis. For example, those areas in England and Wales where people were suffering poor health in the early part of the century (as measured by infant mortality between 1921 and 1925) are the same areas in which adult heart disease rates are high between 1968 and 1978 (figure 4.10); there is a close correlation, both for men (r = 0.69) and for women (r = 0.73). If infant mortality is further divided into neonatal and postneonatal mortality (deaths within the first 28 days, and between one and 12 months, respectively), the association with coronary heart disease (CHD) is stronger with the former. Barker argues that in the areas of high neonatal mortality infants in general were at risk; while the weakest died, others survived into adulthood, but their health was compromised by the poor health and physique of their mothers and poor nutrition. Poor maternal nutrition prejudiced the ability of the mothers to nourish babies in the womb and in early infancy, placing them at risk in later life.

These visual and statistical correlations are only suggestive and not conclusive (moreover, work in Ireland offers only weak support for the hypothesis:

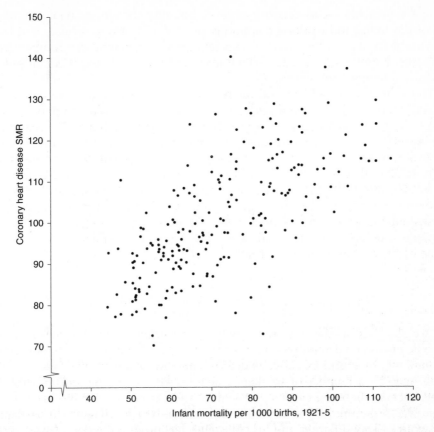

Figure 4.10 Relationship between adult male mortality due to heart disease (1968–78) and infant mortality (1921–5), districts in England and Wales (Source: Barker, D.J.P. (1994) *Mothers, Babies, and Disease in Later Life*, reproduced with kind permission of BMJ Publishing Group)

see Pringle, 1998). One difficulty is that the areas which were poor in the earlier part of the twentieth century remain poor as a new century begins, and so we cannot be certain that the relationship is with early, as opposed to more recent, poor social conditions. When we provide controls for social conditions in adulthood we find that the relationship between infant mortality and adult mortality is weakened, suggesting that geographical variation in mortality is explained by adverse social circumstances in adulthood as much as by early life experiences (Ben-Shlomo and Davey Smith, 1991). What we need as evidence are disaggregated data for individuals rather than data aggregated over areal units. Further, we need some understanding of the biological mechanisms involved. Barker himself provides some from animal experiments, where rats which are undernourished between three and six weeks remain small; as a result, the number of cells in certain body organs is reduced and organ function is correspondingly impaired.

Table 4.7 Coronary heart disease among people born in Hertfordshire, England, in relation to birth weight and weight at one year

Birth weight (lbs)	Men born 1911–30	Women born 1923–30
<5.5	1.00	1.00
6–6.99	0.81	0.87
7–7.99	0.80	0.81
8–8.99	0.74	0.71
9–9.99	0.55	0.52
>10.0	0.65	0.59
Weight at one year (lbs)		
<19	1.00	1.00
19–20.99	0.79	0.59
21–22.99	0.81	0.75
23–24.99	0.62	0.52
25–26.99	0.62	0.84
>27	0.40	0.84

Rate ratios for coronary heart disease: reference category is lowest weight class
Source: Kuh and Ben-Shlomo (1997: 51)

In terms of individual data there is, however, a growing body of evidence that confirms the programming hypothesis. For example, Barker and his team have uncovered historical data collected by midwives working in Hertfordshire, England between 1911 and 1930; such data include birth weight and weight at one year for 37,000 live births. Using the National Health Service Central Register it was possible to trace the subsequent mortality of these children between 1950 and 1992, though it was more difficult to trace women because of changes of name after marriage. Analysis of the relationship between birth weight and CHD mortality is revealing (table 4.7); there is a declining risk of CHD as birth weight increases, in both genders, though the smaller number of women in the sample means that the trend is not significant. There is a similar relationship using weight at one year for males, though again, for females it is less clear. Moreover, these results are confirmed in other countries. For example, a sample of 517 men and women born in Mysore, southern India, between 1934 and 1954, whose birth weight had been recorded, also showed an inverse relationship between birth weight and subsequent heart disease (Stein et al., 1996).

If we accept Barker's findings, the implications for public health are quite profound; namely, to put as many resources as we can into improving the nutrition of expectant mothers. However, some researchers are sceptical about the programming hypothesis. Some dispute the link between early biological "markers" (such as birth weight) and adult disease, arguing that this may simply reflect being born into poor families; the "social" may matter as much, if not more, than the "biological." Others prefer to argue that risks of developing chronic disease cannot be attributed solely to either early life or adult experiences. Rather, such factors (which may be both biological and social) are likely to operate cumulatively throughout life; this is the so-called *life-course* approach to explaining variations in disease incidence (Kuh and Ben-Shlomo, 1997). What evidence favours this mode of explanation?

The programming hypothesis proposes a simple mechanism linking poor development in the womb (and therefore low birth weight) to adult disease. In contrast, a life-course perspective suggests a series of pathways between child-hood and adulthood. These pathways link the parents' economic circumstances to low birth weight and to the child's subsequent socio-economic circumstances; low birth weight may well shape adult disease, but so too do socio-economic conditions throughout life. Further, education may have impacts on health behaviors in adolescence, which will mold later health behaviors and outcomes. For some researchers, then, it is these socio-economic linkages that affect adult health. Family background influences educational opportunity and attainment, which in turn are powerful predictors of adult employment and income; as we have already seen, these impact directly upon health. Powerful support for these ideas comes from longitudinal studies that trace a cohort of people through from birth to adulthood. For example, a National Child Development study was set up in Britain in 1958 to examine a birth cohort; these individuals are now in their forties and their health at this age can now be studied. Power and her col-leagues (1999) constructed a lifetime socio-economic score, based on the father's occupation when an individual was born and at age 16, together with the indi-vidual's own social class at age 23 and 33. There is a striking relationship between the proportion of people reporting poor health at age 33 and this lifetime score.

It is complicated to assess whether the current, or much earlier, stage in the life course impacts upon health. As a further example of this, note how birth weight (taken to represent "early environment") is confounded with obesity (a socially patterned adult risk factor) in terms of its impact on heart disease incidence (Kuh and Ben-Shlomo, 1997: 263). The incidence does indeed increase for those of lower birth weight, but this relationship is mediated by obesity; the gradient is much higher for more obese people and much shallower for those who are not. This, and much other, work indicates that research into social cir-cumstances, and health behaviors, throughout the life course will continue to be productive. It is to such behaviors that we now turn.

Behavioral (Lifestyle) Factors

Here, we consider the argument that people experience poor health because they are more likely to participate in health-damaging behaviors, and that since these behaviors are socially patterned (with those on low incomes or of otherwise lower socio-economic status more likely to engage in unhealthy behaviors) such behaviors are likely to be a key determinant of health inequalities. Typically, the literature deals with four such aspects of lifestyle – diet, alcohol consumption, smoking, and exercise, though we shall also consider, briefly, sexual conduct (table 4.8).

Table 4.8 Health behaviors in England

Social class	Professional and managerial	Skilled non-manual	Skilled manual	Partly skilled and unskilled
Men				
Eats wholemeal bread	21	15	15	14
Eats fruit, vegetables, and salad daily	66	51	59	52
Drinks skimmed or semi-skimmed milk	74	75	65	58
Mean weekly units of alcohol	16.8	15.1	18.9	29.0
Total current cig. smokers	22	30	36	39
Smoking banned at work	35	32	15	18
Smoking not allowed in house	35	32	24	24
Percentage ever using any illegal drug	39	45	33	41
Frequency of exercise (5 days/week or more)	32	35	47	55
Uses sun cream	72	65	55	51
Two or more sexual partners in last year	18	35	25	31
Women				
Eats wholemeal bread	30	25	29	23
Eats fruit, vegetables, and salad daily	81	71	66	68
Drinks skimmed or semi-skimmed milk	84	76	74	64
Mean weekly units of alcohol	7.5	8.0	6.7	6.8
Total current cig. smokers	20	28	35	34
Smoking banned at work	47	43	31	33
Smoking not allowed in house	41	30	25	24
Percentage ever using any illegal drug	29	24	21	25
Frequency of exercise (5 days/week or more)	31	25	39	37
Uses sun cream	84	79	66	64
Two or more sexual partners in last year	13	17	9	17

Figures are percentages unless otherwise specified; social class based on last or current job

Source: Office for National Statistics (Crown Copyright 1997), *Health in England 1996*, reprinted by permission

It is well-established that excess consumption of fats, salt, and unrefined sugars contributes to heart disease. There is evidence from British data that food consumption patterns vary with socio-economic status, those of lower social status groups being less likely to consume fresh fruit and vegetables, foods that have a high dietary fibre content, or milk with a lower fat content. Consumption of alcohol varies across social class for men, but not for women, though the data do not indicate what types of drinks are consumed.

Smoking is a well-established risk factor for much chronic disease, notably lung cancer and heart disease. The class gradients for smoking are clear, for both men and women. Interestingly, this social class difference did not exist 40 years ago. In Britain, smoking behavior depends very much on social circumstances; nearly 70 percent of single parents on low incomes, living in social (public) housing, working in manual occupations, and with few educational qualifications, are regular smokers (Townsend, 1995). The context in which smoking takes places also varies; those in higher social classes tend to tolerate it less in the home, while their workplaces too are more likely to ban it.

Smoking is also implicated as a cause of sudden infant death syndrome ("cot death"). But smoking tends to affect perinatal mortality (stillbirths and deaths in the first week of life) only among women of lower social status, so although smoking may lead to low birth weight (and therefore has a potential impact on perinatal mortality) this does not seem to be true for middle-class women, whose risks of having a poor pregnancy outcome are presumably mitigated by other factors we are considering in this chapter.

As far as other behaviors are concerned, use of illegal drugs and the number of sexual partners seem to vary little according to social class, while exercise is more regularly taken by those in lower social classes, especially men; however, the type of exercise is hidden in these crude figures. Last, the use of suncream (sunblock) is much more common among those of higher social status.

These figures need interpreting with caution, and relate only to one country. Lifestyle or behavioral explanations of health inequalities have attracted plenty of criticism, not least because they have an element of "victim-blaming." The explanation is very individualistic and it is to easy to blame people for their "feckless" or "irrational" behavior. For a young mother seeking to cope with small children in circumstances of severe poverty, smoking may be quite understandable, stress-coping, behavior. Others would wish to see strategies tackling root causes of poverty, or to challenge business interests that seek, via subtle or not-so-subtle means, to capture new smokers, particularly among the young. However, behavioral explanations of inequalities have certainly appealed to governments reluctant to engage in public spending. Consequently, public health strategies have long advocated an emphasis on setting targets for risk factors (reductions in smoking, improvements in diet, increased levels of exercise, and so on) as the chief way of acquiring health gains. But material circumstances play a key role, since the purchase of nutritious and "healthy" food depends upon access to appropriate sources, while income determines whether suncreams are affordable.

To what extent, then, can health behaviors "explain" health variations? Does the social class gradient in smoking, for example, explain the social class gradients in heart disease? On the face of it, the answer to this question should be "yes," since smoking and dietary fat consumption are risk factors for heart disease and are elevated in those from lower socio-economic groups. The White-hall I study, however, showed that cholesterol levels were greater in the higher grade occupations, so it does not appear that cholesterol variations can account for higher rates of heart disease in lower grade occupations. More generally, the "classic" risk factors such as smoking and diet explain only about one-third of the variation in heart disease rates, whether among British civil servants or using international data.

Disentangling behavioral from other factors (particularly the "material" cir-cumstances considered later) is very difficult. This is because although some health-damaging behaviors are more common among groups of lower social status, such groups are also those with poorer access to material resources, such as adequate incomes, decent housing, and secure employment. The general con-clusion concerning lifestyle-based accounts of health inequalities is that they are only a very partial explanation of mortality and morbidity; structural factors and material conditions both shape behavior and have direct impacts upon health.

Social and Community Influences

There is a growing body of opinion that social isolation, and lack of engage-ment in local community life, contribute to poor health and even early death. In a sense, this idea is not new, since the French sociologist Durkheim showed that suicide was more common among socially isolated individuals. More recent epidemiological work suggests that individuals' social networks – the social con-nections and relations they have with friends and family – are a predictor of all-cause, and cause-specific, mortality. A prospective study of over 37,000 American men aged 42–77 years in 1988 (Kawachi et al., 1996) showed that those who were socially isolated (defined as unmarried, having fewer than six relatives or friends, and playing no part in community groups) had an elevated risk of mortality from heart disease, even after allowing for other risk factors. The suggestion from this and similar studies is that social networks provide social support, whether of a tangible (perhaps financial) nature, or in the form of advice and information (perhaps health-related), or more general emotional support in times of stress.

In other work this notion of social support has been extended to the wider concept of *social capital*, which embraces a variety of aspects of the social en-vironment. These include, for example: feelings of trust and safety; the quality of connections among people living in neighborhoods; the number and quality of connections with family and friends; and the extent to which people are members of community groups, societies, clubs, and other organizations. Notwithstanding the difficulties in adequately capturing the concept of social capital, evidence from a number of studies suggests that it has some value in explaining health variations. For example, further work in the United States by Kawachi and others (1997) has measured social capital in 39 states, using data on the membership of voluntary groups and levels of social trust. Low levels of social trust, and of voluntary group membership, are strongly correlated with mortality from most causes, an effect which is only modestly attenuated when an adjustment is made for poverty. In other words, regardless of mean income, states that have poorer stocks of social capital have poorer health. But this is a study of aggregate (and large!) spatial units. To what extent do these findings apply at an individual level?

A study of local-scale health inequalities in parts of north-west England (Gatrell et al., 2000; 2001) found that while loneliness, diet, smoking, and income were significant predictors of self-reported health, some measures of social capital, such as connections with family and friends, and participation in the local community, also contributed explanatory power. Quantitative findings were supported by in-depth interviews. For example, a woman from a deprived area spoke of "more crime in this area that wouldn't have occurred even five to six years ago. . . . car crime, thieving of cars, people abandoning cars and setting fire to them". She indicated that "I find it hard to relate to some people from this area" (Gatrell et al., 2000: 165).

Despite the growing empirical support for social capital as a partial explanation of health divides, it has not gone unchallenged. Some of the variables that comprise indices of social capital are somewhat suspect. For example, simple membership of social groups (the Ku Klux Klan, or exclusive golf clubs, perhaps?) does not necessarily produce social capital. In addition, from a political standpoint it is quite convenient for liberal governments to ask communities to improve social capital, since it diverts attention from more fundamental issues of poverty. Muntaner and Lynch (1999: 72) go so far as to suggest that this is the "sociological equivalent of 'blaming the victim', where communities, rather than individuals, are held accountable for 'not coming together' or 'being disorganized'." Further, from an empirical point of view evidence suggests that material circumstances (see below) are a more important determinant of health divides than is social capital (Gatrell et al., 2000; 2001; Cooper et al., 1999).

The concept of social capital does not apply only at sub-national scales. We noted earlier the reduction in life expectancy among Russian men (Walberg et al., 1998) and, indeed, the increasing east-west health divide (Watson, 1995). In both these studies the authors draw upon ideas of a declining stock of social capital to explain these changes. In Russia, those regions that have incurred the largest reductions in life expectancy are those in which the socio-economic transition to a market-based economy has been felt most acutely and in which income inequality has risen most dramatically; they have also experienced the greatest increases in crime, taken here to be a proxy for social cohesion.

Dramatic evidence in support of the importance of psycho-social factors comes from suicide statistics for Poland (Watson, 1995: 930). Under state socialism rates among middle-aged men were about 30–35 per 100,000 (figure 4.11). With the emergence of the Solidarity movement in 1981 rates dropped suddenly to less than 25 per 100,000, reflecting perhaps a growing optimism about improvement in political and material conditions. In the later 1980s and beyond,

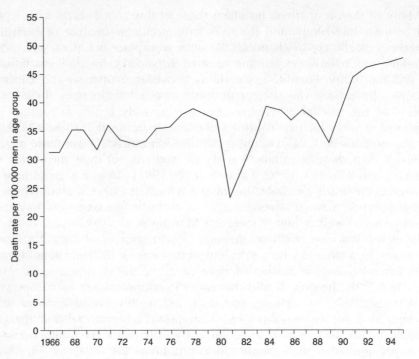

Figure 4.11 Suicide in Poland: men aged 45–54 (Source: Watson, P. (1995) Explaining rising mortality among men in Eastern Europe, *Social Science and Medicine*, 41, 923–34, reproduced with kind permission of Elsevier Science)

rates have returned to pre-Solidarity levels, and even increased, paralleling – and perhaps caused by – a growing sense of despondency about the reforms, a disillusionment with the political system, and a sense of helplessness.

A study that also links community and social factors to wider socio-political processes is that by Wallace (1990). This work deals with violent deaths (homicide and suicide), as well as other health outcome measures, in the Bronx district of New York City; they are seen as "embedded in conditions of pre-existing poverty and overcrowding whose impacts have been exacerbated by the loss of community and of social networks" (Wallace, 1990: 801). The broader context is one of urban decay and removal of basic services, leading to out-migration and the disruption of community relations. But the very high rates of violent deaths (more than one per 1,000 persons in 1978–82) are concentrated in the south-west of the Bronx; only three miles away are some of the wealthiest communities in the country.

Working Conditions and Local Environments

Here, we consider both the "environments" within which people work and those in which they live. Both have major impacts on health outcomes.

Many of the occupations in which those of low social status are employed can be quite hazardous, and the risks from accidents because of exposure to dangerous machinery or chemicals, or poor workplace practices, will certainly go a long way towards explaining some of the social class gradients in mortality and morbidity. Possible associations between occupation and cancer, for example, have been the subject of much research (Alderson, 1986). Other aspects of the workplace environment are possibly health-damaging. Those employed in jobs which are routine and mundane, and where there is little scope for autonomous decision-making ("job decision latitude") are more at risk of ill-health than those who have variety in, and control over, their daily work (Karasek and Theorell, 1990; Karasek et al., 1981). This may go some way to explaining the health gradients among the Whitehall workers, all of whom were engaged in non-manual administrative work but some of whom have higher social status as well as higher incomes (Marmot et al., 1991).

Domestic working environments can also be sources of hazard and stress. For example, a woman who works either part-time or full-time in paid employment often carries the burden of most or all of the domestic tasks. Anyone "managing" the home will find this easier if resources such as an operational washing machine, a telephone, and a car are readily available. Some women working as paid, rather than unpaid, domestic laborers can find themselves exposed to harassment and intimidation by their employers, quite apart from the expectation that they should work long hours for little pay. Clearly, at a micro-scale, the nature of the home environment, whether physical or social, can shape the health of those working, as well as those living, there.

Outside the workplace, in what ways does the nature and quality of the local environment impact upon health? In a classic study, the anthropologist Peter Phillimore has compared Middlesbrough and Sunderland in north-east England (Phillimore, 1993), asking the interesting question: why, when they are so similar in terms of socio-economic mix and status, do they have very different mortality profiles? In the early 1980s, Sunderland had an SMR (all causes, for those aged under 65 years) of 127, while Middlesbrough's was markedly worse (156). Phillimore speculates that the explanation lies in the towns' different industrial environments and histories and in the local geographies of health care provision; Middlesbrough is dominated by petrochemical industries and has a record of poor air pollution, and it has more centralized health care provision, requiring greater distances to be traveled. We return in the next chapter to issues of service provision, but clearly we cannot neglect industrial history or environmental pollution. We need to ask more searching questions about the characteristics of places or "localities;" what they are really like, and what their histories tell us about likely health outcomes. Such histories need, as Phillimore suggests, to consider not merely the physical environment (such as varying air quality), but also local cultures and social environments.

Some of the most significant work along these lines has been done by medical sociologists working in Glasgow, where extensive comparisons have been made of contrasting parts of the city (Macintyre et al., 1993; Sooman and Macintyre, 1995). Like Phillimore, the Glasgow team asks whether there are features of local areas that are health damaging or health promoting. Under this heading

Table 4.9 Local social environment and health outcomes in Glasgow, Scotland

Area indicator	West End (NW)	Garscadden (NW)	Mosspark (SW)	Pollok (SW)
Access to services				
Pharmacy[a]	97.9	96.0	87.2	92.0
Post office[a]	98.4	97.7	89.4	96.4
Grocery store[a]	99.5	99.4	97.9	98.5
Reported problems				
Discarded needles[b]	4.3	8.5	10.8	18.0
Poor public transport[b]	13.4	16.7	20.5	18.8
Assaults and muggings[b]	28.2	34.5	47.9	56.1
Health outcomes				
Health for age[c]	35.1	22.6	29.8	13.2
HADS anxiety[d]	16.5	17.7	29.8	25.6

Notes:
[a] Percentages of respondents reporting amenities within half a mile
[b] Percentage of respondents reporting selected problems
[c] Percentage of respondents reporting health as excellent
[d] Percentage of respondents reporting anxiety

Source: Sooman and Macintyre (1995)

are included such things as access to leisure and recreation facilities, public transport, a clean and safe environment, reasonably priced and "healthy" food, and good health care provision. Even the simplest tallies of resources and facilities suggest that access is worse in the more deprived south-western districts of Mosspark and Pollok (table 4.9); there are higher rates of crime against the person in these areas, and residents there perceive a higher level of "threat" than those in the north-west. In other work (Sooman and Macintyre, 1995) the team focuses more explicitly on residents' assessments of their health, as well as anxiety and depression (the so-called HADS or Hospital Anxiety and Depression Scale). Results for the separate neighborhoods within both the north-west and south-west suggest that the health experience is worse in the latter and this appears to parallel the varying quality of the local environments.

Although we shall consider the health effects of environmental contamination in later chapters, it is worth pointing out here that, since potential sources of pollution are often spatially localized, as is the distribution of low-income groups, it will frequently be the case that the burden of environmental pollution falls disproportionately on the latter. As Schell and Czerwinski (1998) demonstrate via numerous examples, the poorest groups in the USA and New Zealand have the highest exposures to cadmium and to lead, and to dichloro-diphenyl-trichloroethane (DDT) in breast milk, while higher percentages of black and Hispanic Americans live in communities close to landfill sites receiving hazardous waste. The third National Health and Nutrition Survey in the USA revealed that the percentage of children under five years with blood lead

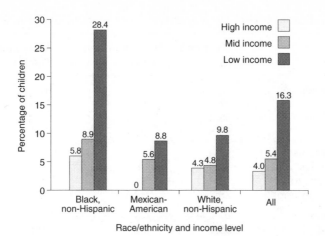

Figure 4.12 Children under five years of age with high blood-lead levels (Source: Schell, L.M. and Czerwinski, S.A. (1998) Environmental health, social inequality and biological differences, in Strickland, S.S. and Shetty, P.S. (eds.) (1998) *Human Biology and Social Inequality*, reproduced with kind permission of Cambridge University Press, Cambridge)

levels in excess of 10 µg/dL varied markedly by both income and race/ethnicity (figure 4.12). Black children in poor families were seven times more likely to have elevated blood-lead levels than white children in high-income families. Such high levels cause cognitive and behavioral impairment, leading to poor educational performance and a possible perpetuation of low income and social status. Environmental risk, and its likely health consequences, is thus socially patterned and serves to cement health inequalities.

This lack of "environmental equity," or environmental justice, is emerging as an important area of research, particularly in North America. For example, the US Environmental Protection Agency has a register of sites that release toxic chemicals into the environment. Such facilities are overwhelmingly located in neighborhoods having high concentrations of those from ethnic minorities and on low incomes (Neumann et al., 1998).

Such work, whether dealing with access to health-promoting "goods," such as recreation and public transport, or proximity to health-damaging features of the environment, is important, since it suggests that health "gains" can come about as much through interventions that are not directly health-related. Perhaps if we spent more on improving the infrastructure of local areas, and in improving local environments, we would secure real health benefits.

Material Deprivation and Health

As we have seen in earlier sections, there is some evidence to link poor health with a lack of services, amenities, and resources in local neighborhoods, as well

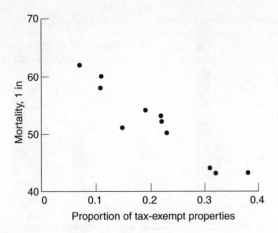

Figure 4.13 Relationship between mortality rates and tax-exempt properties in districts of Paris, 1817–21 (derived from data in Macintyre, 1998)

as a poor physical environment. We have also suggested that a lack of participation in society and a limited set of social relations are also implicated in poor health. The first of these factors leads to *material deprivation*, while the second can be characterized as *social deprivation* (Townsend et al., 1988: 36). In this section we concentrate on material deprivation, recognizing that it also embraces lack of income and wealth, poor quality housing, and unemployment (whether real or threatened).

A concern with health and deprivation is not new. The French writer Villermé (Macintyre, 1998) presented evidence for Parisian districts in the early-nineteenth century, the results revealing a striking relationship between death rates and poverty (as measured by the proportion of properties exempt from taxation because of poverty) (see figure 4.13).

Over the past 20 years a large number of studies have examined how material deprivation has varied by area. This is done by using a set of indicators (usually taken from population censuses) for either quite large or (often) quite small, areal units; these areas might represent populations of perhaps a few hundred people. The indicators are combined, either in a simple additive way, or perhaps using more complex multivariate statistical methods (such as principal components or factor analysis). This then provides a deprivation "score" for an area, to which aggregated health data may be related. Such indices are now widely available in many developed countries; for example, Britain (Morris and Carstairs, 1991; Senior, 1991), New Zealand (Salmond et al., 1998) and Spain (Benach and Yasui, 1999). In Britain, there are four deprivation scores commonly used by researchers and policy-makers (table 4.10 and box 4.3).

To what extent does area deprivation explain variations in health outcome? We can answer this question with reference to a handful of studies drawn from a potentially vast set. These studies, mostly British, relate to a variety of health

Table 4.10 Components and weighting of four deprivation scores

Census variable	Carstairs	DoE	Jarman	Townsend
Unemployment	1	2	3.34	1
Overcrowding	1	1	2.88	1
No car	1	–	–	1
Not owner occupied	–	–	–	1
Lone pensioners	–	2	6.62	–
Lone parents	–	2	3.01	–
New Commonwealth	–	1	2.50	–
Lacking amenities	–	1	–	–
Children under 5yrs	–	–	4.64	–
Low social class	1	–	3.74	–
One year migrants	–	–	2.68	–

The precise definitions of variables differ between indices. See, for example, Senior (1991). DoE is the abbreviation for the Department of the Environment (currently, Department of the Environment, Transport and Regions)

BOX 4.3   *Deprivation scores in Britain*

Although it is quite straightforward to use census data to derive single-variable descriptions of area characteristics (such as the proportion of households owning two or more cars) it is now common to turn to one of a number of multivariate measures of "deprivation." The components of these are spelt out in table 4.10. Two, the "Townsend" and "Carstairs" scores, are simple, unweighted sums of four variables, although the Carstairs index treats "social class" as a component while Townsend uses overcrowding instead; otherwise, the variables are thought to be reasonable measures of material conditions. The other indices have a broader set of variables, including data on ethnicity (New Commonwealth residents) and elderly people living alone, but these and other variables are not in themselves measures of deprivation. Indeed, the Jarman index was originally devised as a measure of the workload of General Practitioners (family physicians) and although it is often referred to as an "underprivileged area" score most researchers use Townsend or Carstairs as a deprivation index.

Some health researchers use data from sources in addition to population censuses, in order to derive other multivariate area descriptions. Such data include those from consumer panels, market research, and credit ratings. These "geodemographic" scores were first used in marketing, but have been adopted by public health specialists. For example, the SuperProfiles system in Britain produces 10 lifestyle groups (such as "affluent professionals", "better off older people") that can be disaggregated into up to 160 cluster groups. These classes can be mapped at the enumeration district scale.

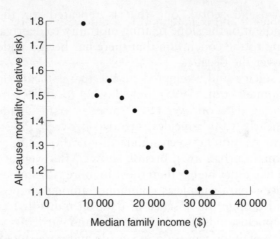

Figure 4.14 Relationship between all-cause mortality and median family income among white American men (derived from data in Macintyre, 1998)

outcomes, including all-cause mortality, suicide, child dental health, asthma, and low birth weight.

Examining data for England, Drever and Whitehead (1995) have shown that when local authorities are divided into quintiles there is a very clear link between all-cause mortality and deprivation. This is true for both men and women, and for all ages. The ratio of SMRs in the most deprived, to those of the least deprived, quintile approaches 1.5. Data on life expectancy (Raleigh and Kiri, 1997) reveal the same picture. Similar mortality gradients are observed in Scotland (McLoone and Boddy, 1994). In the USA, the availability of income data from the census means that multivariate deprivation scores are less widely used; instead, data on median family income by area of residence (zipcode) may be used to portray a very striking gradient of mortality according to this direct measure of deprivation (figure 4.14). These data come from the multiple risk factor intervention trial (known as MRFIT), a prospective study of over 320,000 men screened between 1970 and 1973 (Davey Smith et al., 1996a; 1996b). Adjusting for smoking, blood pressure, and cholesterol did not remove these associations. This has been confirmed by other studies.

Elsewhere in Britain, Senior and his colleagues have examined the relationship between mortality and deprivation in Wales, focusing particularly on how this relationship changed between the early 1980s and early 1990s (Senior et al., 1998). The areal units are electoral wards (the smallest units of local government in England and Wales, typically comprising populations of under 5,000). The map of all-cause SMRs (1981–3) for those aged under 65 years, ignoring small areas where there are fewer than 20 deaths, reveals high levels of premature mortality in the industrial south and north-east of Wales, in some of the poor inner-city and suburban districts of the major urban areas (Cardiff, Swansea, Newport, and Wrexham), and in some small rural towns and seaside resorts. There is a highly significant, linear relationship between mortality and deprivation (Townsend score), one that is repeated for the 1991–3 data. However, the gradient of the slope relating mortality to deprivation is higher in the 1990s, leading to the conclusion that there has been a widening of mortality differentials over the decade.

In a study of suicide and attempted suicide (para-suicide) in part of south-west England, Gunnell et al. (1995) looked at 24 localities (groupings of electoral wards). Taking data on over 6,000 cases of para-suicide between 1990 and 1994, and nearly 1,000 suicides recorded between 1982 and 1991, they show a clear linear relation between standardized ratios and deprivation. Inner-city parts of the main urban area, Bristol, show SMRs well over twice the district average and five times higher than those in more rural areas. Despite gender and age differences (para-suicide is commoner among young women, suicide among men), and despite the absence of data on individual histories and circumstances, the conclusion is that deprivation – and particularly unemployment – is a key factor explaining variation in suicide and para-suicide. At a national scale Whitley and colleagues (1999) have shown that social "fragmentation" (as measured by levels of mobility, single person households, and those renting in the private sector) is at least as important as deprivation in accounting for spatial variation in suicide. This indicates that both material and social factors are important in predicting suicide.

The work on associations between area deprivation and mortality has been conducted primarily in Britain, though there also have been studies in other parts of northern Europe. Recent work in Spain (Benach and Yasui, 1999) indicates that the same associations are found there too. Here, the worst mortality and the most deprived areas are in the south and south-west of the country (Andalusia, for example), while the north-east (Navarre, Catalonia, Rioja, and Aragon, for example) are relatively less deprived and have better health. In parts of the south there may be as many as 8,500 extra deaths per year, relative to the less deprived areas; this is a considerable mortality "gap." Research in Australia (Turrell and Mengersen, 2000) shows that infant mortality is highest in areas of multiple deprivation (as reflected by low income and unemployment, for example); it is also elevated in small areas with high proportions of Aborigines, but it is socio-economic circumstances that are the underlying determinant. However, there is much here to be explained, since while the city of Brisbane has the highest SMR for infant mortality (148) and Adelaide the lowest (97) the cities are similar in terms of the overall proportion of low income and Aboriginal families; these broad city differences remain unexplained.

Morbidity data are linked to deprivation too. For instance, dental decay among children has been shown in a number of studies to be associated with socio-economic deprivation. Conventionally, this is measured by a "DMF" index, representing the number of teeth which are decayed (D), missing (M), or filled (F). Jones et al. (1997) have examined the mean DMF score for a number of small areas (electoral wards) in three districts of northern England, and plot the relation between mean score and deprivation (Jarman score). There is a clear linear relationship (figure 4.15). However, the three districts are characterized by different types of water quality. One district, Salford/Trafford has no fluoride in the water supply; another, Newcastle/North Tyneside, has had the water

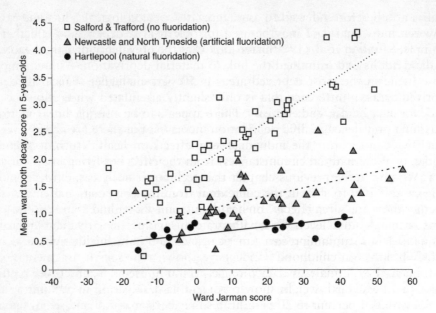

Figure 4.15 Relationship between tooth decay and "deprivation" by small area in three districts in England (Source: Jones, C.M., Taylor, G.O., Whittle, J.G., Evans, D., and Trotter, D.P. (1997) Water fluoridation, tooth decay in 5 year olds, and social deprivation measured by the Jarman score: analysis of data from British dental surveys, *British Medical Journal*, 315, 514–17, reproduced with kind permission of BMJ Publishing Group)

artificially fluoridated, while the third, Hartlepool, has naturally high levels of fluoride. As the graphs show, the relation between dental health status and deprivation varies between the three areas. Children living in the less deprived areas show much less variation in DMF score than those in very deprived areas, suggesting that the latter would benefit in particular from fluoridation. In Britain at least, fluoridation is a politically charged topic and there are strong lobbies arguing that artificially fluoridating water infringes personal liberties (see pp. 224–5 below).

Research linking morbidity (and mortality) to material circumstances in the United States does not have to rely on deprivation indices. This is because, unlike in Britain, household income data are available from the census. We looked above at the work of Durkin et al. (1994) on child injury in Manhattan. Regression analysis of small area data shows clear links between incidence and low household income. Children living in areas with high numbers of low-income households have, on average, twice the risk of severe injury compared with those in higher-income areas; this risk rises to 3.4 for gunshot injury and 4.5 for serious assault.

All these studies are conducted at the area, not the individual level. Data for individuals – on their cause of death, dental health, birth weight, or other measure of morbidity – are aggregated to form a count for an areal unit; such studies are therefore referred to as "aggregate" or "ecological." In some cases, however, individual data may be used to look at area deprivation effects. For example, Salmond et al. (1999) have looked at the prevalence of asthma in New Zealand adults and examined the link to material deprivation. Asthma among those living in the most deprived areas is 50 percent higher than in the least deprived areas, a difference that is only slightly attenuated when allowance is made for age, gender, and ethnicity. There appears to be a simple linear increase in asthma prevalence as area deprivation increases. But there are other ways in which we can separate the individual-level effects on health (such as age and gender, or even material circumstances) from the effects of living in a deprived area. We saw in the previous chapter that techniques such as multi-level modeling can be used to sift out the contextual from the compositional, and to see whether there are "area effects" on health outcomes, over and above those operating at an individual level. We can illustrate this with specific reference to health inequalities by examining recent American studies on low birth weight, and one British example on childhood accidents.

Roberts (1997) analyses data for over 130,000 births in Chicago in 1990, and rates of low birth weight show considerable spatial variation (figure 4.16), from less than 1 percent to 20 percent in some community areas. Given detailed individual data Roberts introduces controls for known risk factors such as age, parity (number of previous children), and level of pre-natal care. This is important, since otherwise we would not know the extent to which high rates of low birth weight simply reflected a preponderance of young mothers, for example (known to be more likely to give birth to low-birth-weight babies) in particular areas. At the individual level maternal race/ethnicity accounts for a considerable amount of variation, with African-American mothers more likely to have a low-birth-weight infant. At the community level an index of economic hardship (combining measures of unemployment and poverty) is associated with high numbers of low-birth-weight babies. Roberts suggests that "women in high-poverty, high-unemployment communities have fewer material resources and therefore run higher risks for malnutrition, lower quality health services, and stress" (Roberts, 1997: 600). He also indicates that social support networks are likely to be more fractured in such areas.

In a parallel study of Baltimore, O'Campo and her colleagues (1997) use multi-level modeling to bring out the area or neighborhood effects. They are thus able more explicitly to answer the question: "Are neighborhood-level variables directly related to an increase or decrease in risk of low birthweight?" As with Roberts' study, individual-level variables such as later initiation of pre-natal (ante-natal) care contribute to the risk of low birth weight. But at the neighborhood level those living in census tracts where per capita income was under $8,000 were at significantly higher risk than women living in census tracts with per capita income exceeding this figure. Interestingly, there is evidence that the effect of individual-level variables varies between neighborhoods, and according to neighborhood characteristics, as shown in figure 4.17. For example, while women with low levels of education are more likely to have low-birth-weight infants, the effect of this is stronger in high-crime rather than low-crime neighbourhoods. On the other hand, in wealthier neighborhoods, where

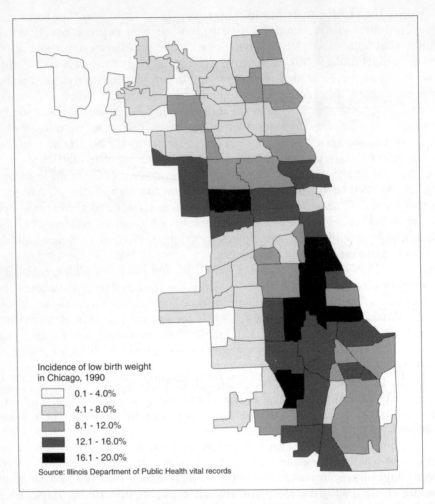

Incidence of low birth weight
in Chicago, 1990

☐ 0.1 - 4.0%

▨ 4.1 - 8.0%

▨ 8.1 - 12.0%

▨ 12.1 - 16.0%

■ 16.1 - 20.0%

Source: Illinois Department of Public Health vital records

Figure 4.16 Incidence of low birth weight in Chicago, by community area, 1990 (Source: Roberts, E.M. (1997) Neighborhood social environments and the distribution of low birthweight in Chicago, *American Journal of Public Health*, 87, 597–603, reproduced with kind permission of the American Public Health Association)

unemployment rates are lower, the benefits of early ante-natal care seem to be greater than in poorer neighborhoods. In any event, this kind of study (see also Gorman, 1999) indicates that explanations, and also health policies, need to target areas as well as individuals. A similar structural argument has been used by McLafferty and Tempalski (1995) in their study of the changing spatial distribution of low birth weight in New York City. Here, women's "vulnerability to economic restructuring, shifts in insurance coverage, and the reconfiguration of health services" (McLafferty and Tempalski, 1995: 320–1) helps explain why increases in the incidence of low birth weight are spatially patterned. It is impossible to separate rates of low birth weight from the changing geography of socioeconomic status.

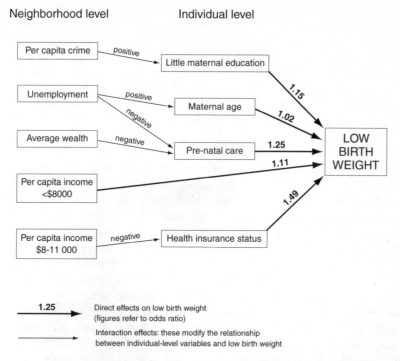

Figure 4.17 Factors affecting risk of low birth weight in Baltimore, 1985–9 (Source: O'Campo, P., Xue, X., Wang, M.-C., and Caughy, M. (1997) Neighborhood risk factors for low birthweight in Baltimore: a multilevel analysis, *American Journal of Public Health*, 87, 1113–18, reproduced with kind permission of American Public Health Association)

As a final example, research on childhood accidents in part of eastern England (Reading et al., 1999) has demonstrated that factors operate to shape risk at both the individual level and that of the wider locality. Although the bulk of variation in injury rates is accounted for by individual "risk factors" (being male, having a single parent at home, having a large number of siblings, for example), living in a deprived area added predictive power to the model. Here, the causal mechanisms seem more plausible than in many studies, since better neighborhood housing conditions, the provision of safe play spaces, and traffic-calming measures are all "area-based" measures that will contribute to lower accident rates.

There is, then, quite convincing evidence for the operation of "place" effects, though there is continuing debate as to whether area deprivation matters less than individual deprivation. The general weight of evidence suggests that there are "place" effects on health, though what matters most is the level of poverty in the individual household (Shaw et al., 1999).

Much of this work, however, provides good *descriptions*, but is less good at offering *explanations* of health inequalities. What is it about "material deprivation" that contributes to poor health? In terms of markers of poor housing (such as overcrowding) then high-density occupation may encourage more rapid spread of infections. Damp housing too will produce fungal spores that lead to poor respiratory health, while lead-based paint or water pipes are further potential causes of morbidity. Low incomes have major indirect consequences for health, limiting dietary choices; it is well known that a good diet is important, but the absence of this from the household may not reflect lack of knowledge so much as poor access to a range of healthy foods, or a restricted choice. Unemployment too will constrain household budgets and have direct health effects, such as psycho-social stress and depression.

For Wilkinson (1996), the causes of health inequalities lie in the deeper underlying structural inequalities within societies. This theoretical position may be used to account for health variations at a number of spatial scales. For example, among developed countries there is only a weak correlation between life expectancy and average incomes. However, there is a much more striking association between life expectancy and the distribution of income within countries; those countries where wealth is more evenly distributed are those with higher life expectancy (Wilkinson, 1996: 72–6). Research both within the United States and Britain confirms this. All-cause (and cause-specific) mortality in US states correlates inversely with the percentage of household income received by the least well-off 50 percent of the population (Kaplan, 1996). In Britain, Ben-Shlomo et al. (1996) have shown that mortality is associated with variations in deprivation within local authorities. Both these studies controlled for average income and deprivation, respectively. At a yet finer spatial scale, Boyle and colleagues (1999) have shown that the variation in deprivation within quite small localities in England and Wales (electoral wards and those adjacent to them) predicts self-reported long-term illness. This adds a further spatial dimension to the literature on associations between ill-health and deprivation. So too does other research in the USA. Fang and his colleagues (1998) show that, in New York City, whites living in predominantly white neighborhoods have significantly lower mortality than whites living in predominantly black neighborhoods. Similarly, black people living in black neighborhoods had lower mortality than those living in white areas (table 4.11). These interesting and important results hold even after controlling for poverty. Membership of a neighborhood majority seems to confer a health advantage.

Watson (1995) adopts an explanatory framework similar to Wilkinson's, rejecting the idea that behavioral factors, the quality of health care, or levels of environmental pollution, play a significant part in worsening health outcomes in eastern Europe. She too stresses the importance of "relative deprivation;" the raising of the Iron Curtain meant that those living in countries behind it had raised expectations of improved living standards that began to approach those in the west. For her, the health crisis is "the outcome of socialism's (ultimately unsuccessful) struggle to modernize in an increasingly global context" (Watson, 1995: 928).

Wilkinson's work, and the further research it has spawned, argues for changes to broader social structures, and ultimately a redistribution of material resources. Many of those interested in the health of ethnic minorities empha-

Table 4.11   Mortality in New York City (1988–94)

	Males		Females	
	Whites	Blacks	Whites	Blacks
In "white" areas				
All-cause	1 474	2 578	910	1 258
CVD	739	979	494	624
In "black" areas				
All-cause	1 934	2 109	1 415	1 177
CVD	982	789	805	561

Age-adjusted mortality rates (per 100,000) for population aged 25 years and over. CVD is all cardiovascular disease; "white" areas are those zipcodes where at least 75 percent of the population is white; similarly for "black"

Source: Fang et al. (1998: 472)

size these structural factors, suggesting that overt and covert discrimination – whether in housing or employment – leads to social exclusion. Risk factors such as those representing health behaviors are seen as merely surface causes of ill-health; the deeper causes are those lying in racism and, more broadly, a lack of access to material wealth (Nazroo, 1998).

An historical perspective on this is given in work on the incidence and spread of tuberculosis (TB) in South Africa (Packard, 1989). Some contemporary explanations of the growing burden of TB among urban black Africans in the early-nineteenth century tended to blame the victims for their poor adjustment to urban living conditions, their ignorance of basic hygiene, and even their adoption of European clothing. Others recognized the lack of proper housing conditions and diet but chastised those infected, as if they chose these conditions rather than having them forced upon them as a powerless workforce. The response to the disease burden was to enforce sanitary segregation. A series of health acts forced the removal of thousands of Africans and "coloreds" to quickly constructed ghettos on the edges of large cities such as Cape Town, Johannesburg, and Port Elizabeth. This had little impact on the TB problem, however: "slum clearance did not mean slum removal in the sense of eradication but simply the physical transfer of slum conditions beyond the city limits" (Packard, 1989: 53). An inadequate supply of decent housing, poor wages, and high rents encouraged overcrowding and increased the likelihood of developing TB.

In later years, the rise of the Nationalist Party after the Second World War saw that the "natural home" of black Africans should be the homelands. This displaced the disease from the main urban centres into "rural dumping grounds" and continued the structural policy of "dealing with TB through the application of exclusionary social controls" (Packard, 1989: 252). Although there have been fluctuations, disease notifications among black Africans and colored groups have been, on average, at least six times higher than among whites. Poor quality housing, poor sanitation, and the long journeys to work on overcrowded trains and buses, all provided continued fertile conditions for the spread of TB infection. Mustering a convincing structuralist argument, Packard argues forcefully that state legislation, a basic neglect of housing, and especially the apartheid policy of creating independent homelands, meant that the black TB problem was methodically removed from the white view.

Marks and Andersson (1990) have looked at the structural determinants of violent death, as well as psycho-social illness, in South Africa. This study pre-dates the formation of a new ANC government (and the presidency of Nelson Mandela), and paints a picture of the culture of violence whose determinants lay in the (former) apartheid regime. Apartheid ensured that the black popula-tion was dominated by white economic power and interests. Echoing Packard, they suggest that the "psycho-social stresses of the labour market have been multiplied by a network of laws which has perpetuated migrant labor and excluded blacks, especially Africans, not only from the central decision-making bodies of the state, but also from effective control of their lives" (Marks and Andersson, 1990: 30). Such migrant labor, police control, and state repression served to dehumanize the individual, and the effects were represented by high rates of violent crime, motor vehicle accidents, alcoholism, and suicide. Data from the annual reports of Medical Officers of Health in four cities support these assertions (table 4.12). Deaths from homicide in 1984–6 among whites represented less than 1 percent of all deaths; yet between 8 and 20 percent of deaths among blacks were murders. Death rates from motor accidents were four times higher among blacks in some places, despite the fact that car ownership was far lower than in the white population.

Table 4.12 Mortality among white and black South Africans (1984–6)[a]

Cause	City	White South Africans	Black South Africans
Homicide	Cape Town	5	137
	Pietermaritzburg	3	41
	Durban	5	0
	Port Elizabeth	9	75
Motor vehicle accidents	Cape Town	18	61
	Pietermaritzburg	12	15
	Durban	22	0
	Port Elizabeth	18	72
Other accidents	Cape Town	13	35
	Pietermaritzburg	5	7
	Durban	15	3
	Port Elizabeth	16	72

[a] Mortality rate per 100,000 people
Source: Marks and Andersson (1990)

Concluding Remarks

The "causes" of health inequalities are the subject of intense on-going debate. While earlier work pointed to the importance of *either* structuralist *or* behavioral (lifestyle) explanations, more recent work suggests that other factors operate. Of these, one's social position throughout the life course merits attention, as do psycho-social factors. The idea of social capital has generated considerable research interest in recent years and does seem an additional determinant of health status. Yet many of the factors interact in complex ways.

We have seen evidence in this chapter of the differing approaches to explanation considered in the first two chapters. Much of the classical epidemiological literature is distinctly positivist in conception, trying to disentangle the relative effects of particular variables whilst providing controls or adjusting for the effects of others. A much more modest literature seeks to hear the individual voices of those on whom health divides have an impact, and calls for a more social interactionist perspective. Other very persuasive work adopts an implicitly structuralist perspective, arguing that health divides are rooted in deep social and economic structures; here, it is poverty and income inequality that cause health inequalities (Shaw et al., 1999).

But are poor health outcomes also a function of variable provision of health services? As far as child dental health is concerned, for example, is this simply a function of behavior (parents not getting children to brush their teeth, or letting them eat too many sweets), a function of wider structural factors (such as lack of fluoridation), or a function of poor provision: an inadequate supply of dentists, or the cost of accessing such care? Are high rates of heart disease in part dependent upon access to services for prevention and treatment, as well as a lack of income, or social support, or poor health behaviors? We therefore need to consider one further possible explanation for inequalities in health outcomes, namely, the varying provision or organization of health services, and the differential use of such services. To what extent do place-to-place variations in morbidity and mortality reflect differences in the provision and organization of care? We therefore examine, in the next chapter, inequity in the provision and use of health services.

FURTHER READING

A good overview of health inequalities from a geographical perspective is provided by Curtis and Taket (1996, Chapter 4).

Comprehensive and up-to-date overviews of health inequalities, of both a theoretical and empirical nature, may be found in various of the essays collected together in Bartley et al. (1998). In particular, the chapter by Curtis and Jones (1998) is an excellent starting point for the geographer, while some useful cross-country comparisons are made in Macintyre (1998). Similarly, there are several good reviews in Marmot and Wilkinson (1999); see in particular the chapter by Shaw and others that covers some of the same material reviewed here. The book by Shaw and her colleagues (1999) is by far the best overview of work on health inequalities in Britain.

Kunst's (1997) monograph has a wealth of comparative material for west Europe and the USA. A special issue of *Social Science and Medicine* (1990, 31:3) considered health inequalities between several European countries, though the coverage is uneven and the material now a little dated. For those interested in the British context (and especially England and Wales) the volume edited by Drever and Whitehead (1997) is a mine of information. You should also consult the Independent Inquiry into Inequalities in Health (1998), particularly for a discussion of policy initiatives, which have not been addressed here. See too the collection of papers edited by Benzeval et al. (1995).

Summaries of the rich vein of work on "programming" may be found in Barker (1992; 1994). For a comprehensive overview of the life-course perspective see the collection of review papers edited by Kuh and Ben-Shlomo (1997).

Much of the literature on health inequalities finds its way into a relatively small number of journals. Of these, *Social Science and Medicine, American Journal of Public Health*, and the *Journal of Epidemiology and Community Health* have been very prominent.

CHAPTER 5

INEQUALITIES IN THE PROVISION AND UTILIZATION OF HEALTH SERVICES

I want in this chapter to turn from inequalities in health outcomes – variations in mortality and morbidity – to look instead at inequalities in terms of the supply and usage of health services. In examining supply, I am concerned with where services are located and how they are configured; in short, how they are *provided*. We shall look at this at a variety of spatial scales, from international contrasts in provision through to variations within countries and local provision. But I want also to discuss how people use such services and whether such use is shaped by location, including the costs of overcoming travel distance to such facilities; service *utilization* is therefore a second key theme running through this chapter. We cannot escape broader issues, however, including how the "need" for services is determined and how decisions about resource allocation are made.

In examining patterns of inequality of health outcome in the last chapter I drew upon evidence from a variety of countries. I shall do so again in the present chapter. I do not present here a comprehensive picture of the way health care services are organized and administered in different countries; this would be a daunting task, not least because new governments have a habit of reforming health care systems soon after they take office! The essential distinction is between those systems that give precedence to the "market," versus those that make provision available, at least in principle, on the basis of need rather than willingness or ability to pay. Curtis and Taket (1996: 107) refer to the second ideology as "collectivist," in contrast to the "anticollectivist," market-led ideology. The latter expects the health care user to pay at the point of use, or to buy into private health insurance that meets the costs of treatment. In a collectivist scheme, health services are funded via income from taxes or compulsory insurance.

Issues in the Delivery of Health Services

Levels of Health Care Provision

As outlined briefly in Chapter 1, it is conventional to separate out different levels of health care provision, usually into three categories. First, *primary* health care is provided in the home, or in a clinic or health centre. Such care could include basic medical attention (from a general practitioner or nurse attached to a health center in Britain, or from a family physician in North America, for example), with a range of services offered, such as prescribing of medication, childhood immunization, and screening for some diseases; it therefore covers many aspects of preventive medicine and health promotion. Primary care also includes other health services, such as dental and ophthalmic care and facilities for the dispensing of medication (such as community pharmacies). All such services will need to be offered locally, since they will be accessed quite frequently.

Secondary care may be thought of as that offered in hospital settings, where patients may be admitted for treatment that cannot take place in a health center or clinic. The route for accessing such care may, as in the UK, be via the primary health care system, so that a doctor or physician in the community refers a patient to hospital for further investigation, diagnosis, and treatment, including more complex surgery beyond the minor procedures that might be performed in the primary care setting. In essence, the secondary sector offers more specialized care and will often be less concerned with prevention and more with cure. If patients require further specialist care they may be referred to a *tertiary* center that will have facilities not available in smaller hospitals. For example, cancer patients may be referred to tertiary centers for investigation and surgery, while those in need of specialist heart surgery would also have these procedures performed in such centers. The boundaries between these three levels are often blurred; nonetheless, the distinction is a convenient one that will be used below. Of course, health care is offered in other settings, such as nursing homes and, for those with terminal illnesses, hospices. Health care often takes place outside the clinic and the hospital.

Geographies of Rationing

Provision is inevitably tied up with issues of resource allocation, priority-setting, or "rationing" of services. We might all like to have a well-staffed primary care clinic in our immediate neighbourhood, open at all hours, but the finances available to any health care delivery system rule this out. Government constraints on health care expenditure mean that we cannot provide all the care we might wish to and therefore decisions have to be made about the nature and range of services to provide. These are questions of a geographical, as well as an economic, political, and managerial nature, since such services will have to be provided *somewhere*. Even if such services are mobile, so that provision need not be delivered at a fixed site, issues arise concerning where to locate the supply of vehicles; this arises most obviously in emergency care, where emergency response vehicles such as ambulances must be located somewhere before they are called upon.

The allocation of scarce resources and the rationing of services might be thought of as having little or nothing to do with geography. However, there is massive geographical disparity in the availability and provision of treatment and therapy. For example, health authorities in Britain (see box 5.1) have adopted different stances towards the services they will pay for; some will pay for expensive drug treatments, such as those used in helping patients with multiple sclerosis or haemophilia, for example, while others will not. This has become known as "postcode rationing." It is hoped to phase out this overt unevenness in treatment, which depends simply on area of residence, by the work of the National Institute for Clinical Excellence (NICE), established in 1999, whose role is to determine what treatment is effective (and, by implication, whether it should be funded).

BOX 5.1   *Organization of medical care in Britain and the USA*

Since entire books are written about comparative health care systems it is somewhat daunting to summarize even two of them in a short section! But this box offers some of the essential ingredients.

In Britain, the ideology is fundamentally "collectivist," with a National Health Service (NHS) established in 1948 to provide health care on the basis of need rather than ability to pay. It is funded both by general taxation and taxes on those in employment. People access the health care system usually on the basis of their registration with a general practitioner (GP) who makes referrals, where necessary, to secondary health care (hospitals). The state provides funding to GPs largely on the basis of the numbers of patients registered with them. Health Authorities (HAs) are given budgets by the government to spend on hospital and community health services and are charged with the responsibility of monitoring the health and health needs of the population within their areas. The number and boundaries of HAs have undergone numerous changes in recent years. Hospital services are now organized into "trusts" that provide the health care commissioned by HAs.

The Conservative government established in the early 1990s an "internal market" that formally separated the HAs (who were the "purchasers" of health care) from the hospital sector ("providers"). GPs were given the option of managing their own budgets as "fundholders," though this system was abolished by the Labour administration that was elected in 1997. The ethos of the former government was to cut the costs of health care by introducing an element of competition; fundholding GPs would "purchase" care from where the best value was provided. As of 1999 general practices are now organized into primary care groups (PCGs), many of which are electing to become trusts (PCTs), in the same

continued

way that hospitals are. Depending, therefore, on the time at which authors are writing, there will be reference to internal markets, the "purchasing" of health care, and fundholding; readers should be aware that this ideology and associated language no longer holds.

In the USA the ideology is "anti-collectivist," with funding of health care coming from private health care insurance schemes (arranged by the individual, but generally tied to employment) and from two state schemes (Medicare and Medicaid, introduced in 1965). Medicare provides insurance cover for people aged 65 years and over, while Medicaid provides cover for people on low incomes. Nonetheless, some 33 million people are uninsured (see table 5.7, page 152). Unlike in Britain, individuals are not required to consult a general practitioner for referral to secondary care; they can consult specialists directly. The unevenness in health care provision, and the burgeoning costs of such provision, as well as insurance schemes, led to the creation of health maintenance organizations (HMOs), a form of what is called "managed care." Here, people pay monthly premiums for a comprehensive package of health care to be provided by a contracted group of health practitioners and organizations.

The Clinton administration sought in the 1990s to introduce further state regulation into the provision of health services. It endeavored both to control rising costs, and also to widen health care coverage. Political opposition, as well as pressure from some vested insurance and medical interests, saw the failure of these attempts. Instead, some authors (Whiteis, 1998) claim that a two-tier health care system, comprising those who can afford to pay and those who cannot, will continue to operate. The hospital system is characterized by increasing concentration, and by the formation of conglomerates ("investor-owned hospital systems") that seek profits for shareholders. Competition between health care providers continues to be the norm.

Efficiency and Equity

When making decisions about the provision of health services there are different criteria that can be adopted. First, we can aim to provide services that are *efficient*; by this we mean how to provide services that maximize health benefit while minimizing cost. Second, we can aim to provide services that are *effective*; are the treatments we offer having real benefit, or are we wasting resources on providing services that offer little health gain? Third, we can aim to provide services in an *equitable* way; are services being provided uniformly to the populations they are designed to serve? These are complex issues, and I do not plan here to go into detail. However, from a geographical perspective there are important implications. For example, as a general rule there is a conflict between efficiency and equity in all planning and resource allocation; there are tensions between the wish from "providers" (or funders or managers) to manage resources efficiently and the wish of users to have those resources provided conveniently and equitably. In a locational sense this means a tension between providing one large site or facility (such as a hospital that offers a comprehensive range of services) and providing several smaller facilities that are closer to the population that needs them. The first solution offers economies of scale for the provider, while the other minimizes travel costs for the user. It is not quite so simple, however, since patients might well benefit more from traveling further to see specialists who are located in centers of "excellence." There are, therefore, important debates to be had about the geography of provision, particularly of secondary and tertiary services, since while the spatial concentration of services might seem "unjust" on the surface, the costs of travel might be greatly outweighed by the quality of service provided (box 5.2).

BOX 5.2   *Does size matter?*

As the text suggests, an important question is the extent to which patients benefit from being treated in a larger hospital, one with a high volume of activity (number of procedures and operations), as opposed to a smaller one that is perhaps located closer to their place of residence. For example, do hospitals in which specialists treat larger numbers of women with breast cancer, or which conduct more coronary artery bypass grafts, see better patient outcomes such as improved survival? This issue is explored in a number of chapters in the collection edited by Ferguson and his colleagues (1997). Some of the issues are considered here.

One advantage of larger, more concentrated units is that these are more likely to be centers of clinical expertise and excellence, since the carrying out of larger numbers of procedures permits skills to be developed and refined. Research evidence suggests that outcomes (such as survival) improve as volume of activity increases. But to what extent is this due to the fact that low-volume hospitals might be dealing with more serious illness, older people, or more emergency cases? Here outcomes may be poorer not because of less expert staff but because they are dealing with more severe cases, the prospects for which would be poor anywhere. Few studies adequately provide controls for severity or case-mix.

A further advantage of larger centers is that they can reap economies of scale (size) and scope (using the same facilities for a variety of purposes). In this case, health care costs are lowered, since the costs of staffing and running one large unit of, say, 100 beds will be less than managing 10 units each with 10 beds. Work shows that economies of scale are exploited in hospitals with between 100 and 200 beds, and that hospitals with many more beds may suffer diseconomies. In England, over half the hospitals have more than 300 beds and this suggests that the scope for further concentration is limited.

Set against these benefits are the increased travel costs, and possibly lower rates of utilization, since, as we shall see later, rates of use decline with distance. Because of these factors, and the evidence that economies of scale have already been reaped, Ferguson and his colleagues conclude that there are no compelling grounds for seeking yet further concentration of hospital services, or merging these onto a smaller number of sites.

Equity in the provision of services must not be confused with equality. Geographically, equality suggests that there should be an even distribution of services per head of population. But what matters much more is equality in relation to need, and this is what we understand by equity. To what extent are services being provided to those who need them? Some have gone as far as to argue that an "inverse care law" operates, where those in most need of health care are least likely to find it available (Tudor Hart, 1971). This clearly goes against one of the founding principles of Britain's National Health Service. Much of the present chapter is an attempt to consider the evidence for such an "inverse care" thesis. But before doing so we need to unpack the notion of "need" and its assessment.

The Need for Health Care

The need for health care takes on a very different perspective according to whether one is looking at population requirements in the developed, or developing, world. In the latter, the basic needs for life, such as clean drinking water or an adequate supply of nutritious food, are far from universally met, although they are taken for granted in the developed world. In developed countries "need" takes on a different dimension. But while millions have basic needs met, others do indeed die because their needs are not met. Good examples are where the expertise needed to treat cancer patients is not provided, or where there are insufficient resources devoted to heart surgery.

The gap between need and demand (sometimes referred to as "expressed need") for health care is often a fine one. For example, some people will demand surgery to change their sex, surgery which they argue is needed in order to live a fulfilling life; others would argue that this is not a health "need" and would prefer to see scarce resources devoted to other forms of health care. Some people living with the problems of multiple sclerosis demand the legalization of cannabis, a drug that they argue alleviates their symptoms; for them it is a need. Difficulties, though, arise when health professionals believe a treatment is ineffective, and this conflicts with the contrasting beliefs of lay people.

In Britain in the mid-1970s a Resource Allocation Working Party (RAWP) sought to suggest a mechanism whereby resources for secondary care could be matched to need. Need was assessed according to the age and sex composition of the population in different health regions, but mortality data (SMRs) was also used as a surrogate for ill-health. The underlying principle was that areas of "need" had to be based at least in part on where the burden of ill-health was most significant. As a result, resources were transferred from some parts of the country (primarily the London region) to others, such as the north and northwest (Jones and Moon, 1987: 275–85). Modifications to RAWP have included the use of morbidity data that first appeared in the 1991 census.

With more recent demands for a "primary care led" NHS in Britain there is clearly pressure to ensure that resources for primary care are distributed equitably. This is done at present by using one of the deprivation scores considered in Chapter 4; general practitioners receive extra "deprivation" payments accord-

Health expenditure by region
(% health expenditure world wide, 1990)

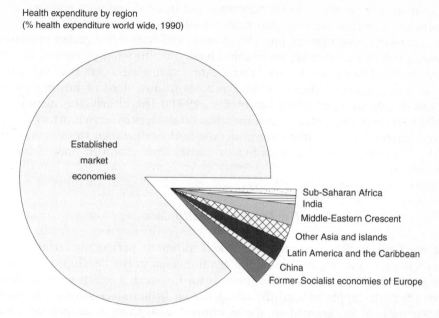

Figure 5.1 Health expenditure by world region, 1990 (Source: Murray, C.J.L. and Lopez, A.D. (1997) Global mortality, disability, and the contribution of risk factors: Global Burden of Disease study, the *Lancet*, 349, 1436–42, reproduced with kind permission of The Lancet Ltd)

ing to the number of patients on their list living in small areas with high deprivation scores. Resources are thereby allocated not simply on the basis of population size, standardized for age, but are "weighted" according to presumed need. This scheme has, however, been the focus of considerable criticism (Senior, 1991; Moore, 1995).

Inequalities in the Provision of Health Services

This section examines the extent to which there is inequity in the provision or availability of health care. It does so with reference to both the developing and the developed worlds, looking at broad variations as well as more local variations in provision.

Health Care Provision in Developing Countries

Recall from the Global Burden of Disease study discussed in Chapter 4 that it was possible to make comparisons between groups of countries (figure 4.1, page 93) according to the common metric of Disability Adjusted Life Years. How is this global burden mirrored, and perhaps ameliorated, by health expenditure (compare figure 5.1 above with figure 4.1)? The answer is that there is an inverse relation, with those countries in "burdened" regions accounting for a much smaller fraction of global health expenditure than is the case in established market economies (Murray and Lopez, 1997). In 1990, about 90 percent of the burden was carried by the developing countries, but they accounted for only 10 percent of global health care expenditure. Sub-Saharan Africa and India together spent less than 2 percent of the global total, but carried 42 percent of the global burden of disease and disability. This variation in expenditure translates into very substantial differences in the supply of doctors or physicians. For example, according to WHO estimates of health personnel, in the sub-Saharan countries of Benin, Chad, Mali, and the Central African Republic there were (in the mid-1990s) fewer than six physicians per 100,000 persons, compared with well over 250 per 100,000 in established market economies.

To what extent do broad differences in health care expenditure translate into the major sectors (primary and non-primary)? In the developing world the distinction noted earlier between primary and secondary care is particularly acute, since the real need for health care is at the primary level, while there is often political pressure to fund prestigious hospital-based developments in urban areas. Put differently, are we to have a health care system with a few highly trained medical professionals, or one where primary care is emphasized, with community workers and traditional medicine playing a part?

Over the past 25 years it has become apparent that "hierarchical," top-down systems emphasizing secondary care have failed to reach the majority. A key date was 1978, when the World Health Organization (WHO) met at Alma Ata in Kazakhstan and declared "Health for All," implementation of which was to be led by primary health workers. Five basic principles underlay this declaration. These were that: a multi-sectoral approach should be adopted; there should be a focus on prevention; local communities should be involved in decision-making; health technologies should be appropriate; and services should be equitably distributed.

In stressing a multi-sectoral approach the reasoning was that primary health care is not only about health; it involves other agencies concerned with education, sanitation, nutrition, and so on. Improved nutrition will not only reduce malnutrition but will also have consequences for later growth and adult diseases (as we saw in Chapter 4). Likewise, clean water will not only prevent dysentery but will lead to other health improvements. Second, it may be more cost-effective to focus on prevention rather than treatment; there is plenty of evidence from the developed world that health improvements in the nineteenth and early-twentieth centuries were due as much to advances in disease prevention, improved hygiene and nutrition as to medical intervention. Third, decision-making should be based on local community needs assessments rather than the objectives of remote policy-makers (Asthana, 1994). Fourth, services need not be provided by highly trained, technologically oriented professionals, but should be community-based. Modernization, for some, means large-scale, impressive hospitals, while primary health care is "low grade;" nonetheless, it might be more appropriate. Last, inequity or the inverse care law operates as much in the developing as in the developed world, with regional disparities in provision, as well as intra-urban and intra-rural ones; we shall look later at some of the evidence.

These were all broad aims, and a comprehensive strategy espoused at Alma Ata gave way subsequently to a more selective approach, focused largely on maternal and child health, since it was argued that the targeting of scarce resources in key areas such as this would be more cost-effective than developing all-embracing comprehensive health care systems (Wisner, 1988; Asthana, 1994). What sorts of strategies are being adopted to improve maternal and child health? In essence, there are seven: growth monitoring; oral rehydration therapy; breast-feeding; immunization; female education; family spacing (planning); and food supplementation. Collectively, using the first letter of each, these are sometimes referred to under the acronym GOBI-FFF. Some of these are discussed below (see Price, 1994, for a lengthier discussion).

Female illiteracy is as high as 90 percent in some areas, reflecting and perpetuating the poor status of women in many countries. It is important to involve the mother in an educated, participative way in the control of her own fertility, and the health of her children. It is also important to look at the qualitative aspects of provision. For example, in Papua New Guinea there is quantitatively reasonable provision, yet work shows that over 70 percent of interactions between mothers and staff take less than two minutes. If the mother is not unwell, little time is spent in giving health advice. Some primary care workers are disease-oriented rather than looking at the health care needs of the individual. Interaction with health services would benefit if female literacy improved.

As well as regulating population growth, family planning seeks to control the health risks of repeated pregnancies, since there can be risks associated with delivery in poor settings. It is estimated that there may be 500,000 maternal deaths each year in the developing world, particularly in South Asia and sub-Saharan Africa (Turshen, 1999: 19). There are huge variations in maternal mortality rates, and these correlate with the proportion of births attended by a trained health care worker (table 5.1).

Barriers to the adoption of contraception are partly cultural but also structural (in terms of the supply of advice and the contraceptives themselves). To surmount this problem in Thailand, where 60 percent of GPs are in Bangkok and most of the others in urban areas, auxiliary midwives are used to promote family planning; here and elsewhere the distribution of oral contraceptives is often via lay personnel.

Until the 1980s there had been a growth in the use of artificial substitutes for breast milk, since bottle-feeding had been seen and promoted as "status-giving." International companies such as Nestlé had mounted vigorous campaigns to market their products; but bottle-feeding is expensive and such infants are more likely to be malnourished and to contract diarrhoeal infections (since the powdered milk is often diluted with unclean water and the bottles may be unsterilized). More recently, there have been campaigns to promote breast-feeding.

Oral rehydration (a simple solution of salt and sugar) is seen as a low-cost solution to the huge burden of diarrhoeal disease which causes dehydration, the consequences of which are either rapid death or longer-term malnutrition,

Table 5.1 Maternal mortality rates and health care provision in parts of sub-Saharan Africa

Country	Maternal mortality rate (1990)[a]	Percentage of births attended (1990–6)
Angola	1 500	15
Mozambique	1 500	25
Uganda	1 200	38
Zambia	940	51
Tanzania	770	53
Kenya	650	45
Zimbabwe	570	69
Industrialized countries	31	99

[a] Rate per 100,000 live births

Source: Turshen (1999: 20)

leading to a further cycle of illness and death. Diarrhoea causes about one-third of all deaths (about 3.5 million each year) in children under five years. The uptake of oral rehydration therapy (ORT) varies regionally and over time. For example, an initially poor uptake in Africa contrasted with substantial progress in other regions, such as south-east Asia and central America; in Honduras, for example, infant deaths from diarrhoea fell by 40 percent within two years of its introduction. Most recent figures indicate, however, that uptake in sub-Saharan Africa has improved (81 percent uptake during the period 1990–7) and has been better than in regions such as Latin America. One aim is to try to get countries to produce their own ORT sachets; this is beginning to happen in Bangladesh and Nicaragua, for example.

WHO set up an Expanded Programme on Immunization (EPI) in 1974, targeting six diseases (diphtheria, pertussis, tetanus, measles, poliomyelitis, and TB); together with ORT, it is estimated that this saved 1.5 million lives of children aged under five between 1982 and 1987, although many more could be saved each year if these strategies were more widespread. While about 40 percent of the world's children had, on average, received basic immunization by 1987, there were still great regional variations. However, it is easy to under-estimate the organizational and logistical problems involved in delivering such vaccines to remote areas, in ensuring that health service workers in unstable countries are not attacked by those opposing the government, and in making sure that clinics are available and open regularly.

The very selectivity of much primary health care in the developing world has been called into question, not least from a structuralist perspective which argues that effects, and not deeper causes, are being treated. Lives may well be saved, but the children will still be open to the respiratory and water-borne infections that contribute so much to the global burden of disease. Better housing, nutrition, and clean water are required, otherwise the "cured"

Table 5.2 Access to safe water and adequate sanitation in selected countries (1990–6)

Country	Percent population with access to safe water	Percent population with access to adequate sanitation
Afghanistan	12	8
Benin	50	20
Cambodia	36	14
Central African Rep.	38	27
Ethiopia	25	19
Gambia	48	37
Haiti	37	25
Madagascar	34	41
Malawi	37	6
Niger	48	17
Papua New Guinea	28	22
Sierra Leone	34	11
Somalia	31	43
Sudan	50	22
Zambia	27	64

Source: WHO (2000)

children simply become re-infected. As Wisner (1988: 965) has it, are we "to believe that oral rehydration therapy is an acceptable substitute for the clean water which would prevent diarrhoea, to which parent and child have a right?" Comprehensive strategies are therefore still required to counteract the appalling lack of access to basic needs of safe drinking water and adequate sanitation (table 5.2).

Others argue that existing programs are not always targeted at those most in need, and that the ethos is wrong, taking the form of passive hand-outs rather than integrated, participatory, health care. Critics such as Wisner argue that GOBI prioritizes a biomedical model rather than a more appropriate social model of health care. "The effectiveness of local organization is further undermined by the individualizing orientation of GOBI elements and their implied model of disease causation focusing not on social causes but on ignorance and faults in individuals" (Wisner, 1988: 967). He further suggests that poverty is too often seen as "natural," with population growth a cause of poverty rather than a symptom. However, addressing this "cause" via a focus on female education and family spacing, for example, merely stresses the "ignorance" of the women involved and bypasses the broader, and more difficult, socio-structural causes. In addition, a selective approach to maternal and child health ignores other health care needs of women in developing countries. The health risks to women of working both inside and outside the home deserve fuller attention than they have received to date (Lewis and Kieffer, 1994).

Turshen (1999) has pointed to the massive structural adjustments in health care since the mid-1970s, which she attributes to the declining role of the World

Health Organization and the growing hegemony of the World Bank. The World Bank is the major donor to developing countries and sets fiscal targets, the response to which in Third World countries has been to cut social rather than military expenditure and to enlarge the role of the private sector in health care delivery. She further argues that the reduction in health expenditure has had severe impacts on women (see table 5.1, page 144). For example, the introduction of user charges in Congo led to a steep drop in numbers of ante-natal visits and visits to child care clinics (Turshen, 1999: 19).

Among other examples, Turshen points to Zimbabwe as a country whose health care delivery has worsened because of structural readjustment. Zimbabwe was an enthusiastic adopter of the Alma Ata declaration, and despite the geographic concentration of health care resources in the two main cities (Harare and Bulawayo) it was able to deliver adequate access to health services to 80 percent of the rural population; the number of rural health centers rose from 500 in 1980 to more than 1,000 in 1990. The proportion of the health budget devoted to preventive care rose from 8 percent in 1980 to 14 percent in 1983. In 1992, however, the imposition of structural adjustment by the World Bank saw the abolition of 1,200 health and nursing posts and a loss of doctors to more lucrative positions overseas. Fees were introduced – in effect, a way of rationing health care – which had drastic and immediate impacts on health outcomes. Exemptions for the poorest people resulted in only limited improvement. For Turshen (1999: 40), the public sector is in disarray and the growing private sector caters only for those willing and able to pay.

The role played by multi-national drug companies in providing health care in the developing world has been documented by a number of authors, including Doyal (1979), Melrose (1982), and Hartog (1993). While the expenditure on drugs in many African countries is a tiny fraction of that in the developed world, it is the uneven distribution, and the nature of the drugs supplied, that generates concern. Coulson (1982) gives some chilling examples from Tanzania, including the widespread use of anabolic steroids (recommended as a "cure" for malnutrition and weight loss, and to stimulate the appetite among the young) and the use (as an alternative to aspirin) of aminopyrine and dipyrone (used as a last resort among the terminally ill in the USA). The latter killed 630 in Tanzania, yet was freely available, without prescription, in pharmacies. Much of this inappropriate use is due to the activity of drug company representatives, one for every four doctors in that country (compared with a ratio of one to 30 in Britain, for example). In the developed world products are carefully labeled with contra-indications and other warnings, while such potential hazards are often missing from the promotional material available in developing countries. A poorly funded public sector and poorly regulated private market (where prescription drugs are sold across the counter or by unskilled people in markets or on the street) mean high rates of self-medication and possible use of the wrong drugs and dosages. Moreover, the products are often non-essential, as shown in a study (Hartog, 1993) of the drugs marketed by the leading European pharmaceutical companies where, on average only 16 percent of the drugs offered are classified as "essential" (table 5.3).

Table 5.3 Drugs manufactured by the 20 largest European drug companies and groups (1988 data)

Company	Total no. of drugs	Proportion of drugs deemed "essential"
Akzo	86	12.8
Astra	59	39.0
BASF	64	18.8
Bayer	87	26.4
Beecham	132	10.6
Boehr	186	5.4
Ciba Geigy	240	15.0
Farmit	82	14.6
Glaxo	138	18.8
Hoechst	319	18.5
ICI	72	25.0
Merck	97	9.3
Nestlé	54	22.7
Roche	222	15.8
Rhone-Poulenc	236	28.0
Sandoz	214	11.7
Sanofi	440	7.5
Schering	106	18.9
Solvay	80	11.3
Wellcome	107	30.1
Total	3,021	16.0

Source: Hartog (1993: 901)

In Bangladesh, 22 of the products in Glaxo's 1981 product list were "tonic" and vitamin formulations, of which only three were marketed in Britain (Melrose, 1982: 38). Malnutrition is better treated with basic foodstuffs than "tonics" or vitamins, since in these cases people will die from lack of calories long before they run out of a particular vitamin. Throughout the developing world, as Melrose shows in detail, patients are prescribed medicines for treating symptoms, with little emphasis placed on attacking the root causes of health problems. It is therefore hard to deny that the multi-national drug companies control and shape health and health policy in the developing world. The profit motive is the driving force; as one spokesperson for the British drugs industry admitted: "I would be talking rubbish if I were to say that the multinational companies were operating in the less developed countries primarily for the welfare of those countries . . . They are not bishops, they are businessmen" (quoted in Melrose, 1982: 27).

I want finally to say something about regional variation in service provision. Some evidence from India points to marked disparities (Akhtar and Izhar, 1994). Per capita expenditure on health and family welfare in 1986–7 shows that this is low in the densely populated northern states of Uttar Pradesh and Bihar (which have a history of poor resources, and a poor agricultural and industrial base) but much higher in states such as Nagaland, in the north-east. The same picture emerges for staffing levels, with Uttar Pradesh having a ratio of one doctor per 15,600 persons while Chandigarh (in Northern India) has one for every 820. Similarly, there are problems of differential access in urban and rural areas, with 35 percent of the rural population having access to some health care within two km, while 84 percent of those living in urban areas have similar access. In part, this reflects an unwillingness of doctors to work in rural areas (Akhtar and Izhar, 1994).

The "primacy" of settlement structure in many developing countries, whereby population is concentrated in the largest city, means that such cities receive resources over and above what they might expect. For example, the Manila urban region in the Philippines has about 25 percent of the total population, yet 43 percent of the hospital beds. Since political power is concentrated in capital cities the health care needs of those in more remote rural areas are neglected. In Ghana in the 1980s spending on specialist tertiary care amounted to 40 percent of the health budget yet this benefited only 1 percent of the population; primary health care spending was only about 15 percent of the budget. According to Turshen (1999), Sierra Leone, Malawi, Togo, Tanzania, and Liberia still spend over 80 percent of their health budget on hospitals.

Urban bias in the provision of health care is shown in other contexts. For example, in South Africa the provision of X-ray services is highly uneven spatially (Walters et al., 1998). While some districts are relatively well resourced, particularly in the Johannesburg region, others are not. One-quarter of all districts in the country – representing over 7.5 million people – have no X-ray facilities. This spatial inequality parallels that for health services as a whole in South Africa. In Andean Bolivia, Perry and Gesler (2000) have made basic use of GIS to examine the provision of primary health care facilities in a remote mountainous part of the country. Much of the region is only accessible on foot. The authors use global positioning systems to record the locations of small settlements, and then construct circular buffer zones around existing health centers, as well as more complex shapes that represent one-hour walking times to and from such centers. The most inaccessible region, Charazani, has numerous settlements located far from the nearest health center, and the authors show how additional modest investment would ensure that nearly half the population in this remote area would be within one hour's walk; currently, one-fifth of the population have to travel for more than a day to get access to basic primary care.

Health Care Provision in the Developed World

Here I consider some aspects of service provision within the developed world. In particular, I shall look at selected aspects of the provision of primary care facilities in Britain and then examine some issues concerned with the unevenness of provision of secondary care in the United States. This is a very selective picture, but some key themes are highlighted.

In Britain, after the creation of the National Health Service in 1948 the distribution of general practitioners came to match more closely the distribution of population, since restrictions were introduced as to where such GPs could practice. However, the need for primary care is not simply a function of population distribution. Needs are greatest where the burden of disease and illness is high. As a result, some authors suggest that we need to take account of morbidity data (such as that on limiting long-term illness, available from the census) in shaping the provision of GPs. For example, a more discriminating form of needs indicator shows that some rural areas are relatively over-provided with GPs compared with deprived urban areas (Benzeval and Judge, 1996).

As table 5.4 suggests, there have been major changes in the structure of general practice since the early days of the NHS. The proportion of those working alone has shrunk dramatically, while the number working in larger practices has increased, particularly in the last 20 years. Single-handed GPs tend to be older than others and are more likely to be male. Geographically, they are concentrated in urban areas, particularly in inner city parts of London and Birmingham (Lunt et al., 1997).

The number of patients registered with a practice ("list size") varies from one part of the country to another (table 5.5). In the London area (North and South Thames Regions) there are about 2,000 patients, on average, on each GP's list,

Table 5.4   Size of general practices in Britain

Percentage of GPs	1952	1980	1994
Single-handed	43	14	11
Six or more GPs	1	12	24

Source: Lunt et al. (1997) and Department of Health (2000)

Table 5.5   Average list size per general practitioner in Britain (1998)

Health region	Average list size
North Thames	2 000
South Thames	1 952
West Midlands	1 908
North West	1 890
Trent	1 886
Anglia and Oxford	1 817
Northern and Yorkshire	1 800
South and West	1 687
Wales	1 706
Scotland	1 454

Source: Department of Health, 1999

while doctors in the south-west have about 15 percent fewer patients. In Scotland, GPs have less than three-quarters the number of patients of their London colleagues.

While some work has been done on the provision of general practice care in Britain, other authors have looked at different aspects of the provision of primary care. It is important to consider "access" not merely in terms of geographical proximity, but also in terms of the quality of provision. Rogers and her colleagues (1998) have adopted an ethnographic approach to equity of access to, and provision of, community pharmacy services in the UK, couching this within a structurationist framework in which the local physical environment and internal spatial arrangement of the pharmacy shapes the social interaction that takes place there. Simply mapping the distribution tells us nothing about the quality of care (advice) offered in different locations. Using a variety of pharmacy settings in north-west England, the authors recorded and transcribed interactions between staff and patients over the period of a week. Particular interest lay in the comparison of inner-city, small town, and rural localities. The inner-city pharmacy was characterized as a "fortress," encased in iron bars and razor wire, in which health-giving advice was negligible. The main business was dispensing prescriptions for methadone, though it did so in a caring and valuable way. The rural pharmacy was a small and old-fashioned "haven," free from the surveillant eye of security cameras, and operating with a low turnover. Here, advice-giving was much more frequent and the conversations were informal and wide-ranging. One patient asks for advice about his hand and is advised by the pharmacist: "I would go to the doctors with that," to which the patient responds: "I would if I could walk straight in." The pharmacist substitutes for the general practitioner. As the authors note, "inequalities may not simply relate to the use of services but to the more subtle and less tangible aspects related to the type of service provided" (Rogers et al., 1998: 372). Those with the greatest need seem to be receiving an inferior service, another manifestation of the inverse-care law.

There have been several studies by geographers looking at the spatial distribution of health services in the United States. Typically (Shannon and Dever, 1974) these map the provision of facilities as a function of total population distribution; for example, the provision of general physicians per head of population reveals relatively good provision in California and New England, and much poorer provision in states with dispersed populations; see table 5.6 for those states with the highest and lowest rates. But this tradition of essentially positivistic work misses the point that the spatial organization of health care systems cannot be divorced from the political and economic context. This "structuralist" perspective on health care has been promoted by Whiteis (1997; 1998), who sees in the contemporary American health care system an "ascendancy of individualistic, intervention-oriented, corporate-sponsored medicine" (Whiteis, 1997: 230). He points to disinvestment in health care in major urban areas, as corporate interests withdraw health and medical services from unprofitable poor communities and switch their investment to more profitable areas. Thus, health maintenance organizations (HMOs), health care providers that put hospitals, doctors (physicians), and other health services under a single group, are becom-

Table 5.6 Ratio of physicians to total population for selected states in USA (1997)

	State	Rate (per 100,000 residents)
Less than 170 per 100,000	Idaho	150
	Mississippi	156
	Alaska	160
	Oklahoma	166
	Wyoming	167
	Nevada	169
More than 300 per 100,000	Rhode Island	324
	Connecticut	344
	Maryland	362
	New York	375
	Massachusetts	402
	District of Columbia	702

Source: Statistical Abstract of the United States, 1999, Table 197

ing increasingly commercialized and oligopolistic. "Liquid assets at corporate HMOs have risen by annual rates of over 15 per cent since the early '90s; a popular strategy has been to funnel these excess profits into the acquisition of other health plans, thus further consolidating corporate hegemony over expanded market areas" (Whiteis, 1998: 801).

This "corporatization" has spatial consequences, as instanced by McLafferty's (1982) study of hospital closure in New York City. McLafferty looked at 39 hospitals that closed between 1970 and 1981, with particular reference to community hospitals. She showed that those communities having a high proportion of non-white residents were more likely to have hospitals closing, a finding supported by other work (Whiteis, 1998: 800). Closures occurred in areas that had already been designated as areas short of medical personnel; these result in longer travel times, especially where the supply of public transport is poor. In addition, corporate health care providers are reluctant to tolerate patients with HIV and AIDS, or others whose care is likely to be expensive. A further impact of closure programs is the loss of employment opportunities to those formerly working in the hospitals; thus, the income base of the local community is further denuded.

Contrasts in the provision of health care between urban and rural areas in the USA are often extreme. Knapp and Hardwick (2000) show that rural areas are poorly served by primary health care professionals (physicians and dentists). Nationally, the ratio of primary care physicians to population is 95 per 100,000, but this drops to only 4.2 per 100,000 in the most rural areas. Similarly, there are 76 dentists per 100,000 people in the country as a whole, but only 29 per 100,000 in the extreme rural areas. Graham et al. (1995) point to the relative under-provision in rural areas of services for people living with HIV or AIDS. In rural areas, rates of infection are rising, particularly among black American women. But many rural hospitals lack the services and facilities for

Table 5.7 The (health) uninsured in the USA

Percentage of the population without health insurance (1996)	
All persons	15.6
Persons aged 18–24 years	28.9
Persons aged 25–34 years	22.3
Persons aged 35–44 years	16.3
Persons aged 45–64 years	13.7
Hispanic population	33.6
Black population	21.7
White population	14.4
Income <$25,000	24.3
Income $25,000–$49,999	16.6
Income $50,000–$74,999	10.0
Income >$75,000	7.6

Source: Miringoff and Miringoff (1999: p. 94)

comprehensive AIDS care. There is poor access to basic primary care, consultation, diagnosis, early treatment, home care, and other support services. This differential is reflected in survival rates; for example, women in metropolitan Atlanta, Georgia, had significantly better survival than those elsewhere in the state. In part, this reflects the lack of health insurance, but it also reflects the closure of hospitals in rural areas. As Graham and colleagues (1995: 444) report, people living with AIDS "in rural areas near urban centers may obtain adequate care given transportation, but in more remote areas, especially in frontier regions, distance can pose an insurmountable barrier."

Disparities in the availability of health insurance in the USA are extreme (Miringoff and Miringoff, 1999; Carrasquillo et al., 1999). Overall, the percentage of the population without health care coverage rose from 11 percent in 1970 to 16.1 percent in 1997. Despite the existence of Medicaid, designed to provide health insurance for poorer people, a quarter of those on incomes less than $25,000 have no insurance. While 14 percent of the white population is uninsured, 22 percent of the black population, and over one-third of Hispanics, lack health coverage (table 5.7). These inequalities work themselves out in space too; of those living in southern states well over 18 percent are without insurance, compared with under 10 percent in more northern states (figure 5.2).

Finally, we should not think of provision solely in terms of a surgery or hospital or other fixed location; health care can be provided in other ways. For example, "linkworkers" can be employed as bridges between lay people and health professionals, as is commonly done among ethnic minority communities. Mobile services can be used to provide health care, as in the screening of breast cancer, for instance. Volunteers and family members provide health care to

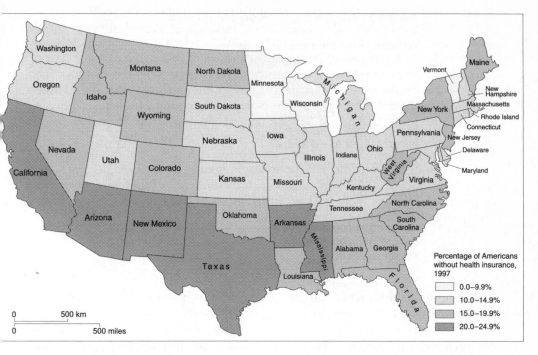

Figure 5.2 Percentage of Americans without health insurance, 1997 (derived from data in Carrasquillo et al., 1999)

Table 5.8 Mean number of hours of care devoted to elderly in Canada, by travel time and gender

Travel time	All respondents	Males	Females
1–30 minutes	4.27	3.55	4.66
31–120 minutes	3.84	2.75	4.63
More than 120 minutes	3.06	2.31	3.74
	p = 0.018	p = 0.029	p = 0.353

Figures are hours per week
Source: Joseph and Hallman (1998: 635)

particular groups, such as the elderly. Even here, distance constrains the ability of some caregivers to devote time to elderly relatives, as Canadian work illustrates (Joseph and Hallman, 1998). Those living less than 30 minutes from their elderly relatives are likely to devote on average 4.27 hours to care per week, compared with only three hours for those traveling for more than two hours (table 5.8). However, this is gendered inasmuch as only male involvement in caregiving declines significantly with increasing travel time.

Increasingly, too, health care is being provided remotely, via what is known as *telemedicine* or telehealthcare. Here, the health professional offers a consultation, or diagnosis, at a distance, using modern telecommunications technology to talk to the patient, or perhaps to view a scanned image of some part of the body, rather than requiring the patient to travel a considerable distance. The applications, both real and potential, of this technology are many and various, but are particularly appropriate for isolated communities where a primary health care professional can get an expert opinion from a major health care center. The technology offers some way of circumventing the problem (see box 5.2, page 139) of wishing to provide specialist care in large centers without disadvantaging those in peripheral regions. Many specialists are now using telemedicine in this way, whether in cardiology, dermatology, obstetrics, or psychiatry, though there is as yet little empirical work on how this improves access to health care.

Utilization of Services

I concentrate in this section on the uptake of health services in the developed world, organizing the material in terms of the level of care sought. We look first at the use of primary care services, before turning attention to the use of secondary and tertiary services. But first some introductory remarks are in order.

In Britain, access to secondary health care (hospitals) is, with the exception of accident and emergency (A&E) services, through the general practitioner. People "consult" their GP, who may or may not "refer" them on to a hospital specialist. To what extent are there place-to-place differences in rates of consultation and referral? Does relative location (distance from clinic or hospital) predict attendance? Is attendance more, or less, likely from those living in relatively deprived areas? If deprivation is a reasonable measure of "need," we should expect higher rates of consultation from more deprived areas, other things being equal. The relationship between utilization, distance, and deprivation or need is complex. Demonstrating that utilization is higher nearer clinics or hospitals may mean that proximity encourages attendance; but it might simply mean that those attending have greater need, because such hospitals may be located in relatively deprived areas. Without adequate data we cannot separate out distance effects from those of deprivation. Distance as a constraint on health-seeking behavior has been studied by geographers and others for many years (see Joseph and Phillips, 1984, for a synthesis of some of this literature).

Use of Primary Health Care Services

For both primary and secondary care there is a frequently demonstrated relationship between service use and area deprivation. As with the study of health outcomes (Chapter 4), an important research question is the extent to which

Table 5.9 Consultation rates in general practice (UK) by distance from surgery

Age group	Distance (km)	Male	Female
Under 5	<2	5.1	4.8
	2–5	4.6	4.5
	>5	4.4	4.1
5–15	<2	2.2	2.5
	2–5	2.0	2.2
	>5	1.9	2.3
15–64	<2	2.5	4.9
	2–5	2.4	4.4
	>5	2.2	4.1
65+	<2	5.2	5.7
	2–5	4.8	5.1
	>5	4.7	5.2

Source: Carr-Hill et al. (1997: 39)

this is mirrored at an individual level. Multi-level modeling can be used to shed some light on this (Carr-Hill et al., 1996). In a survey of consultation rates for 60 practices in England and Wales, data were collected on the socio-economic characteristics of both those consulting and their area of residence. Distance from area of residence (enumeration district) to approximate location of practice was also measured, and an indication of whether the patient lived in a rural area or not was also used. At the individual level, higher rates of consultation were associated with those who were permanently sick, unemployed, living in rented accommodation, of south Asian origin, and living in urban areas, although the magnitude of these effects varied with age and gender. Area characteristics, such as levels of housing tenure and car ownership, had limited additional explanatory power, suggesting that service utilization depends largely on individual factors. Those living closer to the clinics were more likely to consult more frequently, both in urban and rural areas. Results presented elsewhere by Carr-Hill and his colleagues (1997) show that patients living within two km ($1\frac{1}{4}$m) of a surgery are more likely to consult than those beyond that distance, regardless of age and sex (table 5.9). There is thus clear evidence of a "distance decay" relationship between utilization and distance from the health center.

There is other evidence that people living further away from surgeries and clinics consult less than those living closer, even when socio-economic status is allowed for; in other words, it is not because of a lack of need that those living further away consult less frequently. Work in Norfolk, England, by Haynes and Bentham (1982) showed that those living in remoter villages consulted less frequently than others. In rural Vermont, USA, Nemet and Bailey (2000) confirm the inverse association between distance and utilization. Elderly people having to travel more than 10 miles to see a physician are likely to do so much less frequently than those who live nearer their physician. But a more significant predictor of utilization is whether or not the physician is located within a broader "activity space;" that is, the wider set of places that people visit regularly. While the authors' claim that this is linked to "sense of place" (Nemet and Bailey 2000: 1200) is exaggerated there is certainly scope for going beyond simplistic measures of distance in the study of health care use. Evidence from the developing world (for example, Muller et al., 1998) also points to the way in which attendance at primary health care clinics declines markedly with distance.

What explains distance decay effects? It may simply be the friction of distance – the fact that distance, whether in terms of time or cost, deters people from consulting. Alternatively, patients may trade off the "costs" of a consultation against the possible benefits. And when we do consult, what characterizes the consultation? Communication between doctor and patient may be ineffective, either because patients do not understand what they are being told, or because the doctor does not spend sufficient time with the patient. Merely looking at "consultation" or the length of time spent on this, is less useful than knowing how effective (from both points of view) is such health service use. Research indicates that middle-class patients spend more time with GPs than do those from working-class backgrounds and that they also find it easier to communicate with their GPs (Benzeval et al., 1995).

Other work takes a more sophisticated view of "access," seeing this more in terms of possible time-space constraints on service use. Survey research in northwest England (Senior et al., 1993) sought to use a variety of variables to predict immunization uptake. These included accessibility and mobility (travel time to clinic, availability of car), child care commitments, social class, educational attainment of the parent(s), whether or not the parent was single, and whether the child was ill when called for immunization. Statistical analysis showed that educational attainment of both the mother and father, child care commitments, and illness of the child were all significant predictors. In-depth interviews with parents supported these findings and referred to the practical difficulties of attending the clinic: "I'm on my own with three kids and pregnant. I just don't go. I would have to walk or go to the Precinct and get two buses. I have no money for taxis." For another woman, "my husband was at work, and it's a lot of messing taking the other children with me."

Other dimensions of primary care are the use of screening and ante-natal services. How does the uptake of screening for cancer vary from place to place? Two studies of screening for cancer, both based within quite small regions of Britain, are illustrative.

Bentham et al. (1995) looked at screening for cervical cancer in Norfolk. The present UK system requires all eligible women aged 20–64 to be screened every 3–5 years; screening is often done at the general practice surgery (physician clinic). The following variables are significant predictors of uptake: the presence of a female GP; practice size (larger practices have higher uptake rates); and deprivation (the higher the Townsend score, the worse the uptake). However, distance from the surgery has no significant effect; women are not dissuaded from attending by the costs of overcoming distance. In a 1998 study, my

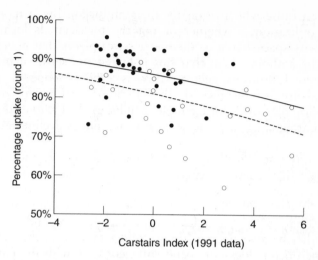

Figure 5.3 Uptake of screening of breast cancer in South Lancashire, UK, 1989–92 (Source: Gatrell et al., 1998)

colleagues and I considered practice variation in uptake of screening for breast cancer in South Lancashire, in north-west England (Gatrell et al., 1998). Although the screening is not done in the practice setting it is possible to determine an uptake rate by knowing which women are registered with particular practices. Uptake is predicted quite well by the deprivation characteristics of the practice catchment area; practices in more deprived areas have lower rates of screening for breast cancer (figure 5.3). The presence of a female practitioner (solid circles in figure 5.3) is also a contributory factor. This suggests that, although not directly involved in the screening, a female doctor perhaps plays a role in giving encouragement to the woman to attend for screening. Whether or not a lack of screening translates into higher morbidity or mortality is an important question that we consider later.

Let us look finally at pre-natal (ante-natal) care. Larson et al. (1997) studied late pre-natal care in the USA between 1985 and 1987. The data they studied relate to over 11 million births, of whom just under 10,000 died before their first birthday; the overall mortality rate was 8.94 per 1,000 live births. Particular interest centres on rurality: specifically, whether residence in a non-metropolitan county has a significant impact on late (or missing) ante-natal care, after adjustment for maternal age, race, education, and other factors. Results indicate that 22 percent of all births are in non-metropolitan areas and that in such areas there is a significant proportion that has had late or no pre-natal care (table 5.10). This is true of all groups except African-Americans. Regardless of area of residence, the proportions of non-whites without adequate care are well over twice those of the white population. Geographically, non-metropolitan residents in most states are at significantly greater risk than urban residents of delaying, or not receiving, ante-natal care. Larson and his colleagues attribute this to the costs of overcoming distance, to declining participation in obstetrics

Table 5.10 Inadequate pre-natal care in the USA, by place of residence and race (1985–7)

	Percentage of total births	Percentage with late or no prenatal care
Whites		
Metropolitan	76.7	4.90
Non-metropolitan	23.3	5.11[a]
African-Americans		
Metropolitan	84.1	10.72
Non-metropolitan	15.9	10.72
American Indians		
Metropolitan	37.5	11.54
Non-metropolitan	62.5	13.83[a]
Other		
Metropolitan	92.5	6.44
Non-metropolitan	7.5	6.87[a]

[a] significant difference ($p < .01$)

Source: Larson et al. (1997: 1750)

by family physicians, and to higher proportions of uninsured women in rural areas.

Use of Secondary and Tertiary Health Care Services

Several studies from different countries have examined the way in which the use of hospital services varies with distance and deprivation. We might expect that distance would be more of a barrier where the health problem is less serious. Conversely, if patients see some prospects of great benefit from their treatment, they are more likely to make use of services. In this case distance may be less of a constraint, although this assumes an ability to pay the costs of overcoming such distance.

Slack and his colleagues (1997) examine rates of hospitalization in the East Midland region of England in an attempt to look at the role of deprivation in access to services. Access is modeled in a more sophisticated way than simply measuring straight line distance between area of residence and hospital. The authors look at the proportion of the population within particular travel times of the hospitals, according to both private and public transport. Results indicate that both deprivation and travel time are significant influences on hospitalization rates; small areas that are more deprived, and closer to the hospitals, are more likely to have higher rates of hospitalization.

Rates of referral for chronic kidney failure (requiring either dialysis or transplantation) in south-west Wales varied with distance to the hospital of treatment (Boyle et al., 1996), although only for patients aged 60 years and over.

This holds even when socio-economic status and ethnicity are controlled for. The explanation for this distance decay effect is uncertain; it may be that general practitioners looking after older patients feel that the difficulties involved in traveling long distances outweigh the benefits of treatment. But in order to ensure equity of access it may be necessary to have further outreach clinics in more local hospitals. In contrast, research in Scotland looking at referrals of patients with testicular cancer to specialist cancer centers (Clarke et al., 1995) found that these were fairly uniform regardless of place; those living in rural areas were as likely to be referred for treatment as those in urban areas.

As with primary care, studying the use of hospital services and access to such services from a geographical perspective requires more than an examination of the effect of distance on use. Let us consider some studies that have explored this broader theme.

One area of tertiary care that has generated some interest is the use of investigative and surgical procedures for treating heart disease, such as angina. There are several such procedures now in use, of which surgery known as coronary artery bypass grafting (CABG) is prominent. Ben-Shlomo and Chaturvedi (1995) have looked at the use of this in north-east London and how it relates to deprivation. Uptake ought broadly to match "need" (here taken to be represented by age-standardized mortality), but results indicate that there is some inequity in relation to need, especially for men. Similar work has been carried out in two health authorities in north-west England, some of which has used qualitative methods to shed light on patterns such as these (Chapple and Gatrell, 1998). This work involved interviews of both the general practitioners looking after patients, and the cardiologists and surgeons to whom they were referred for investigation and surgery. One important finding was the difficulty faced by some patients of South Asian origin, for some of whom there were barriers to adequate care. Those small areas with high concentrations of this group appeared to have very low rates of treatment. As one GP noted, "we perceive that they are not as aggressively treated as the indigenous population. I don't know if that is perceived racism or whether it is actually happening" (Chapple and Gatrell, 1998: 156). In England as a whole other authors (Dong et al., 1998) have shown that women are significantly less likely than men to have cardiac surgery; even allowing for socio-economic factors, smoking behavior, and other illness, the probability of a man having such surgery is nearly three times higher than for a woman. Worryingly, the possibility that this reflects "a more generalized process of doctor discrimination is an uncomfortable notion but is something that requires further investigation" (Dong et al., 1998: 1778).

In Maryland, USA, access to CABG surgery was significantly poorer in patients traveling more than 80 miles (130 km) (Gittelsohn et al., 1995). Other work in the United States has looked at variation in the use of beta-blockers (drugs used in the treatment of heart problems, particularly in reducing the risk of death following a heart attack). Wang and Stafford (1998) show that the these drugs are under-used, notably among patients aged 75 years and over, among non-white patients and among those without private insurance. Clearly, then, there are similar health divides operating here too. Regionally, patients in the north-east are much more likely to be offered beta-blockers than those elsewhere in the country.

Von Reichert and his co-authors (1995) conducted a study of access to obstetric care (maternity services) in rural Montana, with the specific goal of comparing the experiences of white and native Americans. Their interest is in determining whether there are differences between the two groups in terms of travel to access care. However, both groups suffer from rural isolation, since there are relatively few physicians at primary care level and few rural hospitals. The more densely populated counties have hospitals offering a high level of obstetric care, while most counties have smaller hospitals that do not provide specialist care and some have no hospital services at all. The Indian Health Service provides medical care in the form of birthing facilities and clinic services on reservations.

During the study period (1980–9) there were over 130,000 births, of which nearly 90 percent were to white mothers and 10 percent to native Americans. The authors found that while only 19 percent of whites traveled outside their county of residence, 37 percent of native Americans did so. In addition, the latter traveled further to access a lower level of obstetric care. There are no intrinsic differences between the two groups; both prefer to use local services if the quality of provision is high, while both groups will travel if there is a lack of such facilities. In other words, rural residents are at a disadvantage in terms of accessing health care locally. However, the unequal provision of services acts as an additional spatial constraint for the native Americans; since this group is spatially concentrated in sparsely populated areas, and since service provision is also spatially concentrated, they must travel further (figure 5.4). Clearly, innovative health care delivery must be sought, perhaps via mobile clinics and further use of non-medical staff; yet such improvements cannot mask the underlying structural inequality, manifested in part by the ghettoization of the native Americans on reservations.

The use of health services is therefore a function of both their supply and the demand for them. A good example of this comes from work on the spatial variation in abortion rates in the USA (Gober, 1994). Abortion rates have varied considerably from state to state, and over time. Until 1973 there was considerable variation, with most terminations performed in New York and California. Legislation by the Supreme Court in 1973 evened these variations for a few years, although individual states were given greater freedom in the late 1980s to assert legislative control. In 1988 estimated rates varied from over 40 per 1,000 women aged 15–44 years in California, Nevada, Washington D.C., and New York, to under 10 in North and South Dakota. Gober (1988) models this variation using both supply and demand factors. Among the latter, the proportion of the state's population residing in metropolitan areas is a key determinant; the proportion that is black also has a direct and positive relationship with the abortion rate. Other supply-side factors operate indirectly: for example, per capita income influences state funding for abortions, which in turn predicts the rate. Both supply and demand factors therefore need to be taken into account if rates are to be modeled accurately. Of course, this is an aggregate study which cannot take into account some local social movements that have

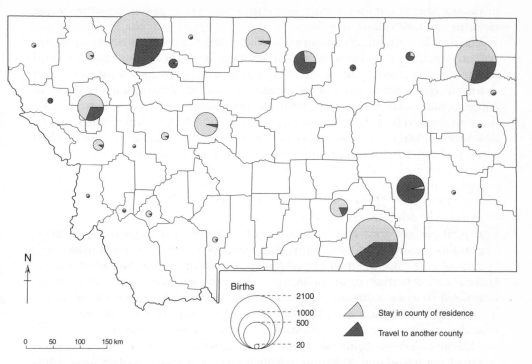

Figure 5.4 Travel for birthing of native American women, 1980–9 (Source: von Reichert, C., McBroom, W.H., Reed, F.W., and Wilson, P.B. (1995) Access to health care and travel for birthing: native American-white differentials in Montana, *Geoforum*, 26, 297–308, reproduced with kind permission of Elsevier Science)

used violent methods to oppose the operation of abortion clinics in some metropolitan areas. Such opposition may well shift utilization from one area to another and cannot be ignored in a full understanding of service use.

Does Provision Affect Outcome?

An important question is whether or not access to, or use of, services has any impact on outcome. If distance does indeed have an effect on health service use, as appears to be the case, are outcomes adversely affected? We can look at the evidence in terms of both emergency and non-emergency care.

Let us look first at outcomes following road traffic accidents (RTAs), involving those travelling in vehicles rather than pedestrians. There is some evidence from the US that fatalities from RTAs are greater in those rural areas that are remote from a hospital or where emergency response times are long. Research in Michigan found that even after adjusting for driver's age and gender, and the characteristics of the crash, the relative risk of fatality was 1.51 in remote rural areas. Bentham (1986) looked at mortality due to RTAs among men aged 15–24 in all local authority areas in Britain; those areas without an Accident and Emergency Department had significantly higher mortality than those which possessed one. This is, however, a very simple measure of provision. More recent work in Norfolk (Jones and Bentham, 1995) suggests that the likelihood of a death from RTA is associated with age (those aged over 60 years have a higher risk) and road speed (death is more likely on roads where speed limits exceed 60 m.p.h.), but, importantly, the ambulance response time does not affect mortality risk. There is an obvious explanation of why this contrasts with American findings, and this concerns the relative travel times to and from the scene of the accident. In more remote parts of the US these times might exceed two hours, whereas average ambulance travel times in England (to the accident scene and thence to hospital) are unlikely to exceed 25 minutes.

BOX 5.3   *Survival analysis*

In survival analysis we seek to describe and explain the risk of occurrence of an outcome (usually death) as a function of time. For example, we might measure survival time from the date of diagnosis to date of death, or the period from date of an operation to date of death; but the outcome might be any other designated type. Here, we assume that death is the outcome. Of course, in any given study a fraction of the population under investigation will not have died at the time of study; such data are referred to as "censored" data.

Survival analysis begins with the construction of a survival curve, which shows the proportion of people surviving beyond a given point in time. An imaginary example is shown in figure 5.5. Initially, all patients will be survivors, but as time progresses a proportion of them fail to survive; as a result, a survival curve is a typically step-like function. "Flatter" curves represent better overall survival than those which drop more rapidly. Of more interest than one such curve, however, is to compare curves corresponding to groups of patients or particular variables that are hypothesized to affect survival. In a non-geographical context this would typically be to compare patients who either have, or do not have, some form of treatment or therapy. But in terms of the focus of attention in this book, we could contrast patients living in different geographical areas, whether deprived or more affluent. For example, Kim et al. (2000) sought to examine whether survival following surgery for colo-rectal cancer was affected by deprivation; do patients in more deprived areas have worse survival rates? Survival curves (figure 5.6) indicate that those in the most deprived 25 percent of electoral wards in a part of southern England have poorer prospects than in less deprived areas. It is important to allow for the effects of other variables, however, in the statistical analysis. For example, it might be that those in the most deprived areas are less likely to come forward to diagnosis than others; they might also be older people. Consequently, survival analysis seeks to fit models (usually *Cox proportional hazards* models) to estimate the separate contributions that different predictor variables make to the analysis.

Survival analysis can be performed using proprietary statistical packages. For basic introductions see McNeil (1996) and Altman, D. (1991) *Practical Statistics for Medical Research* (Chapman and Hall, London).

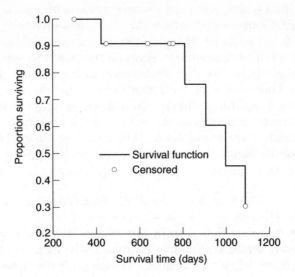

Figure 5.5 A survival curve

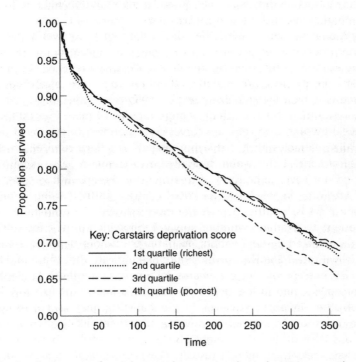

Figure 5.6 Survival curves for colo-rectal cancer patients in Wessex region, southern England, by category of deprivation (Source: Kim, Gatrell and Francis, 2000)

Turning to non-emergency care we might ask whether survival following surgery for cancer varies with location (box 5.3). Research in France found that people with colo-rectal cancer living in rural areas, remote from specialist treatment centers, had poorer survival rates, indicating that there might be delays in getting treatment (Launoy et al., 1992). Subsequent work on colo-rectal cancer in southern England by Kim et al. (2000) showed that there were marked variations in survival following surgery. To a large extent, variation in survival depends upon stage at diagnosis; if the cancer has already spread extensively before surgery the prospects are much worse than if it is not advanced. It is also influenced greatly by age – older people have poorer prospects – and by whether or not the surgery is an emergency operation. Kim and her colleagues (2000) demonstrate that, once these major factors are controlled for, clear geographical differences in survival remain. Survival in some districts is 75 percent worse than in others. The reasons are unclear, but may have to do with the volume and quality of surgery being performed (see box 5.2 above, page 139). This work, like others, shows that simply ranking areas, or hospitals, on the basis of survival rates or other "performance indicators," is fruitless unless it provides controls for case-mix (the type of patient and disease being treated).

Breast screening is of particular interest given the high incidence of breast cancer in developed countries, since poor uptake of screening is likely to be reflected in higher disease rates in future years. American research suggests that women in lower-income groups were less likely to have had a mammogram within the past year; only 9.2 percent of women on incomes less than $10,000 in 1987, compared with 20.9 percent whose income exceeded $35,000 (Wells et al., 1992). This translates into delayed diagnosis, particularly among black and Hispanic women (Richardson et al., 1992), as well as into poorer survival rates – as research from a wide range of countries suggests (Schrijvers and Mackenbach, 1994).

We should not imagine that the links between health outcomes and service provision are only to be assessed in quantitative terms. A good example of how this may be examined qualitatively is a study of women's and community health centers in Adelaide, South Australia (Warin et al., 2000). These centers offer an alternative model of health care to the "fee-for-service" (and more medically focused) primary care that is more commonly provided in Australia. The health practitioners in such centers are concerned with a "social" model of health, and with health promotion in the community. Interviews with users reveal that this model of provision is very much welcomed and that it improves mental health and well-being. As one user put it: "they still have that time to talk to you as a person whereas in general practice . . . . . I feel it's like a factory production line, an assembly line" (quoted in Warin et al., 2000: 1873).

Concluding Remarks

Inevitably, there is geographical variation in the provision and use of health services. The real question is whether this "variation" also represents "inequity;" is there a poor relation between unevenness in provision or uptake of services, and the need for health care? This chapter has sought to examine some of the evidence, which suggests that there is indeed in general a mismatch between "need" and provision. We have also looked at the relationship between service provision and health outcomes.

In the developing world, issues of service provision revolve less around poor access *relative* to others in such countries, and more to whether or not there is adequate *absolute* provision. In other words, less attention has been devoted to whether low-income or deprived groups suffer relative to more wealthy ones. Of more importance is whether there is an adequate overall level of health care delivery (especially primary care). Even so, there is evidence of geographical disparities, with too great a concentration of resources in urban areas.

In the developed world, we have seen that geographical distance constrains access to, and uptake of, health care facilities. But many other factors are important, and "access" to health care in the United States, for example, depends in many cases on the affordability of health insurance. Distance decay effects are commonly observed in the utilization of services, but this is also shaped by whether or not the area of residence is "deprived." More research is needed on whether poor access to services has negative consequences for health outcomes; the evidence on this remains thin.

FURTHER READING

An excellent account of health service delivery, including material on different forms of health care delivery, needs assessment and issues of equity, is provided in Curtis and Taket (1996). Their Chapter 5 goes into more detail than I have concerning different national health care systems. The book also covers the work of the World Health Organization, including its "Health for All" program (Curtis and Taket, 1996: 254–9).

For a good overview of service provision and use in developing countries see the collection of papers edited by Phillips and Verhasselt (1994).

Although now rather out of date, Chapters 6–8 in Jones and Moon (1987) have much of value to say about inequality in the provision of health care, and these chapters, like the rest of the book, merit careful reading. Similarly, the book by Joseph and Phillips (1984) is a classic text that covers in more detail some of the ideas, if not the more recent studies, addressed in this chapter. The collection of papers in Ferguson et al. (1997) is well worth reading.

CHAPTER 6

PEOPLE ON THE MOVE: MIGRATION AND HEALTH

Migration is the permanent or semi-permanent change of residence of an individual or group of individuals. I want in this chapter to consider the links between migration and both health and health care, though as we shall see it is the associations between migration and health outcomes that have been given priority in research to date.

I consider here only the health consequences of human movements that involve a change in residential location. Other forms of human movement dwarf migration in terms of the volume of social interaction and may have potentially dramatic consequences for health and health care. International migrants form only about 1 percent of those traveling as tourists or on business. The widening of travel offers the potential for exposure to diseases not encountered at home. For example, there have been outbreaks of malaria around airports in mid-latitude countries (such as Geneva, in 1989), where infected *Anopheles* mosquitoes have found their way onto airplanes, survived the flight and subsequently escaped to infect local residents (Haggett, 1994: 102). More generally, passengers may be infectious but symptomless while traveling, but the speed of travel means that the trip is over and passengers long dispersed throughout a country before any disease symptoms develop. Another example of the link between tourism and health would be the impact, on the health of both local and visitor populations, of sex tourism, typically involving those from the developed world visiting developing countries. Time-space "convergence" (the shrinking of distances) or "globalization" thus has potentially major public health implications.

Here I focus mainly on permanent, or quasi-permanent, migration. My examples tend to be contemporary, although, as we saw in Chapter 2, migration was linked intimately with colonial history and the spread of disease in the New World and elsewhere. Migration can take place over short distances, perhaps from one part of a town or city to another, but can also be inter-regional, that is, movement from one part of a country to another; and, of course, it can be international.

Migration interacts with health and health care in several ways. First, it may have either short- or long-term consequences for health outcomes. For example, we can consider the short-term impacts of both voluntary and forced population movement on individuals and groups, including those left behind. We can also look at the possible impacts of migration on the spread of disease, and ask whether the incidence of disease in geographical areas is elevated as a direct consequence of population movement. We may consider whether those migrating acquire the health status of those they join rather than those left behind. We might ask whether our understanding of the spatial distribution of disease is incomplete unless we take into account the impact of migration.

Second, we must ask if the health status of migrants differs from those of non-migrants. To what extent are migrants a non-random sample of the population as a whole, in terms of their health status? Is their health status generally better, or worse, than the people who do not move? We shall consider here this issue of health "selection," that is, whether migrants are inherently healthier than those left behind.

Third, we must look at the impact that migration has on the use, or lack of use, of health services. Do people migrate in order to get better access to health and social care? To what extent does the movement of health care professionals impact upon health care provision? How do those providing health care manage the health of migrant groups?

All these are questions that this chapter seeks to answer. Each of these three broad relationships is treated in turn.

Impact of Migration on Health

Migration and Stress

There is evidence that migration, especially to a new country, leads to stress and depression, as a result of alienation and the need to come to terms with a new culture. This is clearly gendered, in that migrant women may have fewer opportunities for social integration, as well as working in unskilled occupations and on low incomes (Kaplan, 1988). One outcome may be the adoption of unhealthy behaviors; for example, Carballo and Siem (1996) refer to high rates of drug abuse among young Puerto Ricans moving to New York.

As one example of the health impacts of the migration experience, consider the movement of Fijian women to British Columbia in Canada. Elliott and Gillie (1998) report on a qualitatively based study set within a social interactionist framework, in which 20 women were interviewed in depth about their health and health status, but also about their migration experience. Some reported problems with back pain as a result of the move, while others referred to problems in adapting to a very different climate, constraints on their mobility, and a struggle to maintain cultural traditions. The search for new work resulted in extreme fatigue for some, while easier access to alcohol for her male partner resulted in the physical and emotional abuse of one woman. More generally, there are problems of loneliness and homesickness, especially for older women. The pressures to perform a variety of roles (wife, carer, mother, homekeeper, wage-earner) are exacerbated by the separation from friends and family and the need to fulfill such roles in an unfamiliar setting.

Table 6.1   Relative risk of suicide (and 95 percent confidence intervals) for Swedish-born and other women (1985–9)

Country of birth	Relative risk in Sweden	Relative risk in country of birth
Sweden	1.0	1.0
Finland	1.68 (1.43–1.98) +	0.92 (0.80–1.05)
Poland	1.63 (1.04–2.56) +	0.48 (0.43–0.53) –
Russia	3.71 (2.05–6.71) +	0.74 (0.69–0.81) –
Germany	1.44 (0.96–2.16)	0.77 (0.71–0.84) –
Hungary	3.39 (2.04–5.63) +	1.84 (1.68–2.03) +
Norway	0.99 (0.63–1.50)	0.57 (0.48–0.68) –

Relative risks are adjusted for age: + denotes significantly elevated, – denotes significantly lower than reference category (Sweden)
Source: Johansson et al. (1997)

Age-adjusted rates of self-reported long-term illness seem to be much higher among foreign-born than indigenous people in Sweden (Sundquist and Johansson, 1997), even after controlling for social and material circumstances, smoking and exercise. The odds of needing treatment for severe long-term illness are over twice as high for both men and women born in southern European countries compared with Swedish-born people. The same picture of worse mental health among migrants to Sweden emerges from research on suicides (Johansson et al., 1997), where women from other countries (notably Russia and Hungary) living in Sweden are at significantly higher risk of suicide than women born in Sweden (table 6.1). The risks are much higher than for women still living in eastern Europe. Johansson suggests this is linked to disrupted social and cultural networks and poor material circumstances, as well as possible discrimination and xenophobia. The social and cultural "distance" from Sweden may simply be too great to overcome, and the consequences are severe.

Immigrants often find themselves exposed to higher risks in the workplace, whether from agricultural and industrial accidents (where those employed illegally will often be working in hazardous conditions) or in domestic environments (where young women in particular may be at risk from physical or sexual abuse). For example, in a comparison of occupation-related accidents among the native-born and immigrants in western Europe, Bollini and Siem (1995) show that rates among the latter are two or three times greater than among the former (table 6.2).

Ugalde (1997) refers to research on agricultural labor migrants to Spain that shows high rates of poisoning from exposure to pesticides. In addition, agricultural firms were ignoring safety regulations and failing to protect their workforce from agricultural hazards. Accidents at work were not reported, in order to mask the fact that labor conditions were "irregular." Ugalde quotes some African immigrants: "There are many who cannot sleep, they think that they

Table 6.2 Accidents among native-born and immigrants (western Europe)

Country	Native-born	Immigrants
Netherlands[a]	32	92
Germany[b]	79	216
Switzerland[c]	158	230
France[d]	11.4	21.5

[a] Rates of occupational accidents per 1,000 insured workers
[b] Rates of occupational accidents per 1,000 workers, all industries
[c] Rates of accidents per 1,000 workers in building trade
[d] Rates of accidents per 1,000 workers in building trade, involving permanent disability or death

Source: Bollini and Siem (1995)

are not going to be given the papers, and if the police stop you. . . . too many problems and you cannot sleep." "If you have papers you are fine, otherwise you are worried. . . . nervous" (Ugalde, 1997: 94). Further, given that employment and incomes are unstable among immigrant laborers, that families are separated, and that racial discrimination occurs within the working environment and beyond, the burden of mental illness among such immigrants is not hard to understand. This suggests that in order to understand health outcomes among migrant communities we need to look at the organization of labor in agriculture and industry and ultimately at wider underlying inequalities in society and the economy.

Migration also has both short- and long-term impacts on the health and welfare of those in the areas from whence the migrants have come. For example, over the short term those migrating may transfer some of their earnings to the families they leave behind and this may mitigate the loss of economically active people from particular areas. Kanaiaupuni and Donato (1999) show how those Mexican communities which have lost population to the United States are areas where infant mortality is high, but that the remittance of income back home ("migradollars") goes some way to offsetting the disruptive economic, if not social, effects of migration.

Other work indicates that, perhaps over the longer term, the migration of more skilled people simply leaves behind a pool of poorer people and serves to reinforce regional disparities in economic and social welfare. Thus, Thomas and Thomas (1999) suggest that in one county in North Carolina (referred to anonymously as "Step County") a rural ghetto emerged over many years as better-educated black people moved out to the central town ("Sparksburg") or out of the county altogether. This reduced the stock of social capital and reinforced residential segregation. The polarization of black and white groups, coupled with the low ratio of males to females (because of the out-migration of many men), may lie behind currently high rates of infection by sexually transmitted diseases. Similar work in the rural Johnson County, Tennessee, hit by closure of manufacturing plants (Glenn et al., 1998) also suggests that over only four years (1990–3) the health of non-migrants worsened, largely as the result of reduced household incomes; at the same time, younger, more affluent people moved into the area as the result of an economic "rebound." This regeneration, and the influx of healthier, wealthier people, masks the problems of the "stayers" who are not participants in the improved economy. Migration then, whether of out-migrants or in-migrants, alters economic and social landscapes and may have long-lasting health consequences for those who have not moved.

The Health of Refugees

The forced dislocation of people from their homes has drastic health consequences, the nature of which dwarfs many of the less serious health outcomes discussed in this chapter. Important though they are, the stresses of the migration event, and the possible subsequent adoption of new disease profiles, hardly bears comparison with the mortality and morbidity wreaked by civil war, political persecution and the "ethnic cleansing" of the late-twentieth century. Forced migration does not only embrace refugees but also includes, for example, those compelled to leave because of major engineering projects, such as dams (Boyle et al., 1998; Roy, 1999). However, for the purpose of this present section we shall consider refugees to be those forced, whether by invasion or domestic conflict, to leave their usual place of residence in order to seek refuge in another country.

Data from the United Nations High Commission on Refugees (UNHCR) may be used to map geographical variation in the size of the refugee population (for example, in Africa; see figure 6.1), though this does of course fluctuate from year to year. Kalipeni and Oppong (1998) have reviewed the health impacts of the displacement of refugees in Africa, considering these effects under several headings. First, these movements disrupt livelihoods, the production of food, and the operation of health services. This results in malnutrition. For example, during the conflict in Somalia in 1992 nearly 75 percent of children under five years of age died, almost all from malnutrition linked to the war. Second, the overcrowding of refugee camps causes further food shortages, as well as poor sanitation; as a consequence, diseases such as cholera, dysentery, hepatitis, and measles are likely to break out. Third, sexual violence is common: UNICEF reported the widespread rape of young girls during the Rwandan genocide of 1994 (box 6.1), while sex may be exchanged for food among the most desperate, an exchange which, like the shortage of uncontaminated blood, does little to halt the spread of HIV across the continent. Fourth, the trauma of watching people, often family members, killed and mutilated has devastating consequences for long-term mental health. UNICEF reported a survey in Angola in 1995 which revealed that two-thirds of children had seen people murdered, and over 90 percent had seen dead bodies. In addition to these health impacts

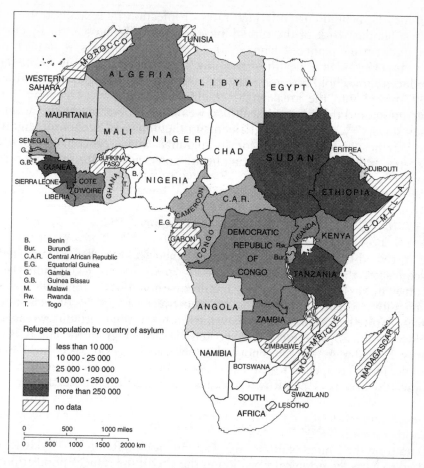

Figure 6.1 Refugee population in Africa, 1998 (derived from data supplied by United Nations High Commission on Refugees)

the chaotic mixing of populations exacerbates the diffusion of newly emerging infections.

Banatvala et al. (1998) studied levels of mortality and morbidity among Hutu refugees from Rwanda who had been repatriated from the Democratic Republic of Congo (formerly Zaire) in late 1996 (box 6.1). The genocide in 1994 had resulted in serious depletion in the numbers of health workers, and a consequent over-stretching of health care delivery. Banatvala and his colleagues paint a graphic picture of the burden of disease and ill-health, including attending to bullet wounds and limb injuries caused by machetes. In earlier work a group of doctors working in Goma, the main town in Zaire into which the refugees moved, estimated a crude mortality rate of 20–35 per 10,000 per day, due mostly to outbreaks of severe diarrhoeal disease (Goma Epidemiology Group, 1995). This contrasts with a crude mortality rate of only 0.6 per 10,000 per day in pre-war Rwanda.

BOX 6.1　*Conflict in Rwanda*

The history of ethnic unrest in Rwanda is a long one but centers on the juxta-position of a majority Hutu population and minority Tutsi population. The latter held sway during the colonial period but after independence in 1962 the Hutus took power. In 1994 the Tutsi-dominated Rwandese Patriotic Front overturned the government, forcing over a million Hutus into exile in North Kivu region, Zaire (now the Democratic Republic of Congo), during July. In the month that followed almost 50,000 refugees died, largely as a result of cholera and dysentery. By December 1995 over 2.9 million refugees had fled their homes. One million had gone to Zaire, half a million to Tanzania, and 1.2 million were internally displaced.

Many refugees entered Zaire via the town of Goma, thousands dying there and others moving on to refugee camps further north. With poor water supply and worse sanitation there were outbreaks of cholera that killed an estimated 50,000 refugees by the end of August. In addition, between 18 and 23 percent of children aged under five years were acutely malnourished, a figure that com-pares with "only" 5–8 percent in non-refugee populations of Africa. The distri-bution of both food and medical supplies was hindered by local Rwandan military and political leaders. As the Goma Epidemiology Group (1995: 343) points out, "the world was simply not prepared for an emergency of this magnitude."

The Impact of Migration on the Spread of Disease

We saw in Chapter 1 how human movement assisted the diffusion of HIV in Uganda. Immigration also had consequences for the spread of tuberculosis (TB) in South Africa during the late-nineteenth century (Packard, 1989: 38–43). Two groups in particular contributed to the spread of disease. One was from England, a group already suffering from "consumption" (the archaic term for TB) and seeking a cure in a warmer climate; often, they infected fellow passengers on board ship, in overcrowded cabins. Upon arrival, some were employed as private tutors and acted as sources of infection among middle-class white families, while other infection arose as a result of local African servants taking home left-over food from consumptive employers. Death rates from TB were socially patterned (see Chapter 4); in one district, Aliwal North, mortal-ity among whites was 20 per 1,000 in 1897, while for black Africans it was approximately 62 per 1,000. A second source of "imported" cases was from eastern Europe, where Polish and Russian Jews were escaping persecution. A high proportion was already infected with TB, and the lack of restriction on movement in some parts of South Africa meant that racially mixed slums grew up that were ideal for disease transmission.

Other associations can be found between migration and TB in South Africa in the late-nineteenth and early-twentieth century, reflecting the role of migrant labor in spreading the disease into rural areas (Packard, 1989, Chapter 4). Several hundred thousand African mine workers moved to and from their work-places and rural homes each year. "This general pattern, in combination with the desire of sick migrant workers to return home and the mining industry's practice of repatriating all diseased or injured miners, guaranteed that the epidemic of urban-based TB quickly spread to the rural hinterlands, infecting the rural households from which industry drew its labor" (Packard, 1989: 92). While reliable data are not widely available, reports and case studies from local medical officers are highly suggestive. One concluded that TB accounted for over 30 percent of notified deaths between 1904 and 1906 in some parts of the Transkei and Siskei "homelands." Disease transmission was aided by over-crowded housing; in some areas there were taxes on the building of huts, thus discouraging a reduction in residential densities.

Tuberculosis remains a major public health problem and continues to be associated with migration. Carballo and Siem (1996) report on studies that show high rates among Haitians moving to the USA, Asian migrants in Australia, and those seeking asylum in Switzerland. But they argue that the high rates are not so much a function of their movement from high-risk areas, as of the poor, overcrowded living conditions in which newly arrived people find themselves. Evidence for this comes from several European countries, including Moroccan migrants to France and Cape Verde Islanders moving to Portugal (Carballo et al., 1998). Poor-quality housing clearly has other health impacts, such as increasing the risk of accidents in the home.

Rates of HIV infection too are influenced by migration, as we saw in Chapter 1. For men moving to work in construction, mineral extraction, and forest exploitation, their contact with sex workers is likely to facilitate the spread of the virus. A good illustration of this comes from migrant laborers in South Africa, where there were nearly three million living away from home in 1986, about 400,000 of whom came from nearby countries such as Lesotho, Mozam-bique, and Malawi. Jochelson et al. (1991) conducted in-depth interviews with a small number of those employed in mining, and this qualitative material reveals mechanisms for coping with separation from families: "After washing my workclothes I go out of the hostel because it's lonely there and there is nothing to while away this loneliness. I go to the likotaseng [domestic worker quarters] to look for women" (quoted in Jochelson et al., 1991: 164). The authors argue that the migrant labor system institutionalizes a geographically based network of relationships for spreading sexually transmitted diseases. A mixed hierarchical-contagious pattern of disease spread is posited: "once HIV enters the heterosexual mining community it will spread into the immediate urban area, to surrounding areas, from urban to rural areas, within the rural areas, and across national boundaries" (Jochelson et al., 1991: 169). But it is the economic system of labor migration that is the deeper cause of disease spread; the behavior of individuals cannot be separated from the vulnerability of family relationships, the low wages paid to the laboring men, and separation from the women left behind. All of these factors are shaped by the underlying socio-economic structures.

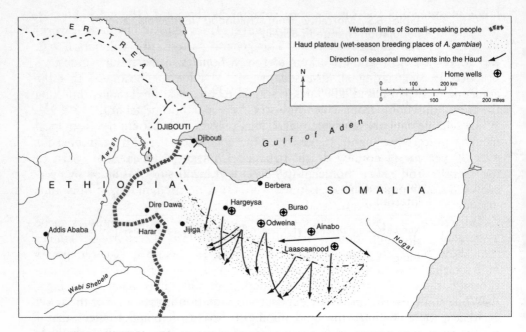

Figure 6.2 Movements of Somali pastoralists in the Horn of Africa (Source: Prothero, R.M. (1977) Disease and mobility: a neglected factor in epidemiology, *International Journal of Epidemiology*, 6, 259–67, reproduced with kind permission of Oxford University Press, Oxford)

As noted in the introduction, population movement need not involve a permanent change of residence, and in some parts of the world more seasonal movements can have major consequences for health. Prothero (1965; 1977) has illustrated this convincingly in his work on malaria during the 1950s and 1960s. For example, Somali pastoralists spend the dry season (November–March) in the north, with access to well water. But the land cannot support cattle throughout the year and they move south to the Haud, a plateau whose hollows fill with water during the wet season and provide ideal breeding grounds for the mosquito vectors of malaria (figure 6.2). The complex patterns of population circulation have created difficulties in devising adequate monitoring and control strategies, especially since this circulation involves crossing the border into neighbouring Ethiopia.

Migration and the Incidence of Disease and Ill-Health

Direct effects

In some cases, migration may have other direct effects on disease incidence. This lies behind one of the possible explanations for "clusters" of childhood leukaemia arising near nuclear installations, such as that at Sellafield in west Cumbria, northern England (see box 3.2, page 66).

Table 6.3 Mortality from childhood leukaemia and population change in Britain

	Males	Females	Total
Population change >50%	37.7	39.1	38.4
Population change <50%	30.2	24.1	27.2
Relative risk (95% confidence interval)	1.25 (0.9–1.7)	1.62 (1.2–2.2)	1.41 (1.1–1.8)

Numbers are mortality rates per million children aged under 15 years

Source: Langford (1991)

While some have argued that exposure to radiation is the cause of the raised incidence of leukaemia, the epidemiologist Leo Kinlen hypothesizes that it results from exposure to an (as yet unidentified) virus, and that an outbreak of infection is most probable when there is an unusually high degree of population mixing. In essentially rural communities, such as that in west Cumbria in the later 1950s, or on the coast of northern Scotland (near the Dounreay nuclear power station), the local population is "challenged" by the rapid influx of a new population, brought in to manage and operate the nuclear installations. The immunity of the local population is low, and rapid immigration leads to a sudden increase in the exposure of the newly mixed populations to viruses. Leukaemia may be a rare response to a viral infection.

To test this hypothesis, Kinlen and his colleagues (1990) studied new towns in Britain, all of which had seen significant population growth after 1950. These towns were grouped into two classes; first, "overspill" towns (such as Basildon, Stevenage, and Welwyn Garden City), designed to provide accommodation for people dispersed from London after the Second World War; second, "rural" new towns (including Glenrothes, Corby and Cwmbran) that took population from a much more diverse set of origins. The hypothesis was that leukaemia incidence in the second set would be higher than among the first. The results bore this out. For example, between 1947 and 1965 the ratio of observed to expected deaths from leukaemia among children aged under five years in the rural new towns was 2.75 (a statistically significant excess), compared with 0.95 in the overspill towns (fewer than expected cases).

Other authors have conducted research that lends support to the Kinlen hypothesis. For example, Langford (1991) looked at deaths from childhood leukaemia (under 15 years of age) in England and Wales between 1969 and 1973, classifying these by geographical area (local authority). Areas are classified according to whether they have undergone substantial or more modest population change. His results (table 6.3) indicate that there are significantly higher death rates (especially among females) in the areas undergoing more population change.

A more sophisticated test of the Kinlen hypothesis requires data on the diversity of migrant origins, since it is the mixing of dissimilar populations that lies at the heart of this hypothesis. Accordingly, Stiller and Boyle (1996) construct a "migration diversity" measure, showing whether migrants tend to come from one source, or many sources. Those districts in England and Wales that had higher rates of migration, as well as more diverse sets of origins, were those in which leukaemia incidence was elevated.

Further work by Kinlen suggests that rural areas with a large influx of oil workers during the 1970s in Scotland are also those with excess rates of child leukaemia (Kinlen et al., 1993). The home addresses of over 17,000 such workers in Scotland were classified as either rural or urban, and the incidence of childhood leukaemia in both groups of areas was compared. Rural areas with the highest proportions of oil workers were those with the highest leukaemia incidence.

There is, then, quite a body of evidence (including that from other countries: see Kinlen and Petridou, 1995) to support the theory that it is migration and population-mixing, rather than direct or indirect exposure to radiation from the nuclear industry, that lies behind leukaemia incidence and leukaemia "clusters" (see also Alexander, 1993). However, it would be very premature to suggest that the case is now closed. Looking at other rare cancers, such as that of the eye (retinoblastoma), it seems that here possible exposure to nuclear radiation may be a causal factor. Morris and his colleagues (1993) recorded cases whose grandfathers had worked at Sellafield, but whose mothers had spent time living near the reprocessing plant. The suggestion was that the mothers may have been exposed as children to radionuclides in the general environment, or contaminated at home, and that they had had an increased rate of cell mutation, passed on to the next generation; thus while the mothers themselves were healthy, their children were not. The children with retinoblastoma had never lived near Sellafield. This illustrates well the point that recording address at diagnosis may have no bearing on the origin of a disease, an issue to which we shall return.

Longer-term health impacts of migration

There is a substantial literature, much of it by epidemiologists, on the differences in health status between migrants and non-migrants, specifically on the extent to which those moving from one area to another (and mostly from one country to another) ultimately "adopt" the health profile and "risk factors" for disease of those who have always lived there. As Elford and Ben-Shlomo (1997: 228) suggest, "migration provides a naturally occurring experiment which may establish the aetiological importance of factors acting at different points in the life course." In particular, it serves to establish the relative role of genetic, as opposed to environmental and socio-economic factors; if migrants from an area of low chronic disease incidence ultimately develop a higher disease incidence in their destination area this would suggest a more modest role for genetic factors.

We may look at cancer and risk factors for heart disease, as well as self-reported health, to illustrate this "acculturation" hypothesis. For example, English South Asians (those whose ethnic origin is in India, Pakistan, or Bangladesh) show higher rates of lung cancer and breast cancer than on the Indian sub-continent, results that are consistent with the notion of a transition of low cancer risk in the country of origin, to a higher risk in the country of residence. Winter and colleagues (1999) suggest this is due to changes in lifestyle among the migrants. Much of this work derives from a classical epidemiological tradition in which "ethnicity" is under-theorized, usually in terms of categories (such as "Indian") that may mask more than they reveal. As Elliott and Gillie (1998: 329) argue: "most studies serve to aggregate ethnic groups to such a level that individual differences in health and/or life beliefs become obfuscated in the name of adequate sample sizes; explanation and prediction take privilege over understanding." It is worth asking whether alternative explanatory frameworks might add to the existing literature.

The question arises as to the length of time which it takes for health status to be modified. A study of rural migrants to Nairobi, Kenya, found significant increases in blood pressure (box 6.2), as well as body weight, only 10 months after migration (Elford and Ben-Shlomo, 1997: 234). Blood-lead levels among pregnant immigrants to South Central Los Angeles were significantly higher than among pregnant non-migrants (2.3 µg/dl, compared with 1.9), but this depended strongly on length of time since immigration. The greater the period women had been in the United States, the lower the blood-lead level. In Canada, obesity among immigrants increases with length of residence (Cairney and Ostbye, 1999). Obesity is usually measured using the BMI index (weight divided by the square of height; a body mass index of 25kg/m^2 is the upper limit of the "normal" range). Women born in Canada have a mean BMI of 24.9 but while those living in Canada for less than five years have a mean BMI of 22.7 this rises to 24.8 for those living there for more than 10 years. This acculturation effect remains even after adjusting for socio-economic and lifestyle factors. Porsch-Oezcueruemez et al. (1999) studied Turkish immigrants to Germany and found that total cholesterol levels were broadly similar to those in other western countries but very much higher than among those living in Turkey. In addition, blood pressure among men, and obesity in women, were also high in the immigrant population.

This kind of adaptation does not apply solely to those moving from one country to another. For example, the British Regional Heart Study has compared the blood pressure of non-migrants, born in the town where their blood pressure was examined, with the blood pressure of migrants. Men moving from southern England to Scotland had higher mean blood pressure than those staying behind (an average of 157/89, compared with 142/80), while those born in Scotland and moving south had lower mean blood pressure than those remaining in Scotland (143/80, compared with 148/85) (Elford and Ben-Shlomo, 1997: 234–5). (See box 6.2 for some information on blood pressure.)

BOX 6.2   *Blood pressure*

Since many studies of migration focus on changing blood pressure it is worth explaining briefly what blood pressure is and what is meant by *hypertension*.

Blood pressure is the pressure exerted on the artery walls by contraction of the heart's lower chambers (ventricles). Peak pressure is known as *systolic* pressure, while that between heart beats is *diastolic* pressure. Blood pressure is measured in millimetres of mercury. A normal reading would be approximately 120 mmHg (systolic) and 80 (diastolic), referred to as 120/80.

Hypertension is abnormally high blood pressure and is potentially dangerous in that it can lead to heart disease and stroke (box 4.1, page 94). Body mass, consumption of alcohol, and intake of potassium are thought to influence high blood pressure. In middle age an increase in systolic pressure of 18 mmHg and 10 mmHg in diastolic pressure doubles the risk of stroke and increases the risk of heart disease by 50 percent. Regular monitoring of blood pressure, and reduction of high blood pressure, is therefore essential.

See Whincup and Cook (1997)

Research in central Africa has compared the growth of Hutu children whose families had migrated from Rwanda to Zaire to work in the copper mines, with that of non-migrants (Little and Baker 1988). In Rwanda itself, the dominant Tutsis tended to be taller and heavier than the subordinate Hutus, but the migrant Hutus were taller and heavier than those remaining in Rwanda (figure 6.3). This was perhaps due to improved hygiene, diet, and health care in the Katanga province of Zaire.

In a series of studies, epidemiologists have looked at the health of those living in the Tokelau Islands, a series of atolls in the South Pacific, north-east of Fiji. Many islanders migrated to New Zealand after a hurricane in 1966 caused much damage, and studies of the health of migrants and non-migrants have focused on blood pressure, as well as chronic disease such as diabetes and asthma (Little and Baker, 1998; Elford and Ben-Shlomo, 1997: 229–30). In the mid-1970s, 812 adults were examined in Tokelau (532 non-migrants) and New Zealand (280 migrants); all had been examined five years earlier, before the migrants had moved. At that stage, there were no significant differences in blood pressure, but after adjustment for age and other factors the blood pressure of the migrants had risen significantly. Such longitudinal work has clear advantages over studies that simply look at migrants and non-migrants at one fixed point in time. Further work has indicated that Tokelauan children living in New Zealand had significantly higher blood pressure than children still living in the islands, even after taking into account weight and height, while other evidence suggests raised incidence of diabetes (especially among women) and asthma in the migrant group (figure 6.4, page 180).

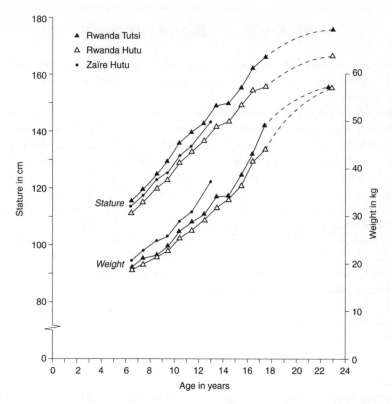

Figure 6.3 Growth in stature and weight of sedentary Rwanda Tutsi and Hutu, and Hutu children whose families had migrated to Zaire (Source: Little, M.A. and Baker, P.T. (1988) Migration and adaptation, in Mascie-Taylor, C.G.N. and Lasker, G.W. (eds.) *Biological Aspects of Human Migration*, reproduced with kind permission of Cambridge University Press, Cambridge)

Research on Samoan populations endorses the notion of migrants "assuming" the health status of those living in the destination area. For example, men living in Western Samoa have a mean body weight of 73 kg (161 lbs), while those in American Samoa (Manu'a and Tutuila) tend to be heavier (about 83 kg, or 183 lbs). However, American Samoans moving to Hawaii and California are heavier still (mean weight of about 88 kg (194 lbs) and 114 kg (251 lbs) respectively). Both men and women in California are markedly heavier than non-migrants. Again, this is reflected in higher rates of diabetes and heart disease (Little and Baker, 1988).

I think it is fair to suggest that much of this literature is of a descriptive nature rather than offering convincing explanations of differences between migrants and non-migrants. The implication in many studies is that migrants simply adopt "lifestyles" (particularly diets) of the populations among whom they come to live. But the extent to which changing (often reduced) material circumstances, or poorer stocks of social capital, play a role has yet to be fully explored.

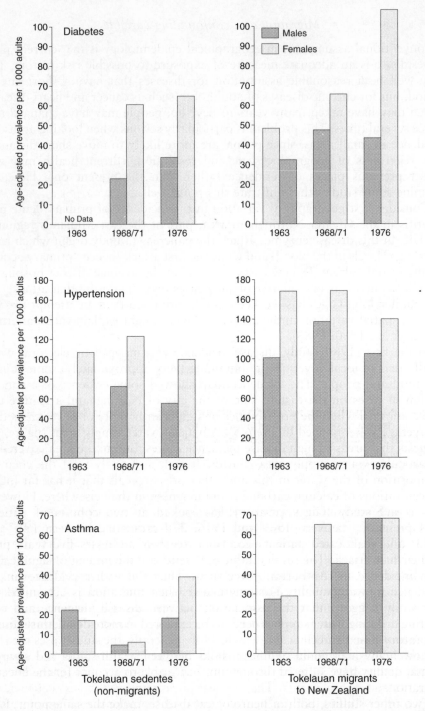

Figure 6.4 Rates of morbidity among Tokelau islanders and Tokelauan migrants to New Zealand (Source: Little, M.A. and Baker, P.T. (1988) Migration and adaptation, in Mascie-Taylor, C.G.N. and Lasker, G.W. (eds.) *Biological Aspects of Human Migration*, reproduced with kind permission of Cambridge University Press, Cambridge)

Migration as a confounding variable

A conventional assumption in geographical epidemiology is that current place of residence is an adequate measure of exposure to possible risk factors. This may well be a reasonable assumption for diseases that have a short latency period, but for chronic diseases of adulthood (such as cancer and heart disease), which may have taken many years to develop, people may have changed residence several times. This problem is particularly serious when looking at disease incidence in small areas, since people are more likely to move short distances. But, regardless of geographic scale and resolution, current health may well reflect previous places of residence rather than the present one. Here, we examine some studies that illustrate this point well.

Consider first geographical variation in the incidence of primary acute pancreatitis in the Greater Nottingham area, part of the east Midlands region of the UK. In this disease enzymes attack the pancreas (a body organ which regulates sugar levels in the blood) and while the first attack (hence "primary acute") results in only about 26 new cases per year in the average district hospital in the UK, as many as one quarter of such cases may die during that attack. The geographer John Giggs has worked closely with doctors in an attempt to see if there is spatial variation in incidence, and what might explain such patterning (Giggs et al., 1980; 1988).

In the first (1980) study, disease incidence was mapped by electoral ward, small areas of local government containing (very approximately) some four to five thousand people. The incidence map showed some evidence of a concentration of cases on the eastern side of the city of Nottingham, a feature that was confirmed when the probabilities of significantly more or fewer cases than on average were mapped (figure 6.5). A map of water supply areas (figure 6.6) suggests that areas of high disease incidence correspond to the area covered by a particular water supply area (Burton Joyce), and analysis of the chemical composition of the water in this and other areas reveals that it has far higher concentrations of calcium carbonate and magnesium than elsewhere. However, the second study (Giggs et al., 1988) looked at two cohorts of patients (214 presenting between 1969 and 1976; 279 presenting between 1977 and 1983) and reallocated patients according to their addresses five years prior to their first attack. The results (table 6.4) reinforce the finding of significantly high incidence in the Burton Joyce area. Thus the earlier evidence linking incidence to water quality is strengthened when migration is accounted for. The study suggests that the association between disease incidence and contrasting drinking water supplies needs to be explored in more detail; this requires laboratory-based and clinical research. More generally, the study raises the issue of how many marginally significant links between health status and environmental quality have suffered through not being able to account for the effect of migration.

Two other studies, both of neurological disease, make the same point. Riise and his colleagues (1991) studied 381 people who had developed multiple sclerosis (MS) in the Norwegian county of Hordaland between 1953 and 1987. They obtained data on place (community) of residence at birth, and at various

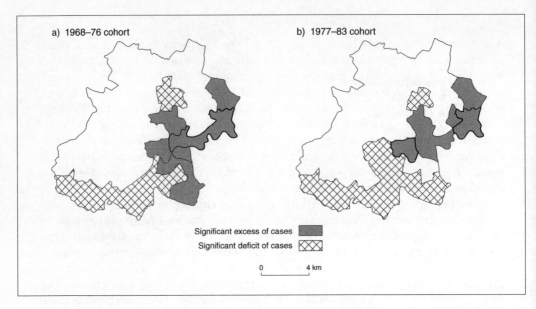

Figure 6.5 Areas of significantly elevated, and significantly lower, primary acute pancreatitis in Nottingham, 1968–83 (Source: Giggs, J.A., Bourke, J.B., and Katschinski, B. (1988) The epidemiology of primary acute pancreatitis in Greater Nottingham: 1969–1983, *Social Science and Medicine*, 26, 79–89, reproduced with kind permission of Elsevier Science)

Table 6.4 Primary acute pancreatitis in Nottingham, England

Water supply area	CaCO₃ (ml/l)	Average annual incidence rates (per million)			
		All cases		By former address	
		1969–76	1977–83	1969–76	1977–83
Derwent	197	62.8	92.2	62.8	82.4
Redhill	196	66.7	129.2	64.5	85.2
Oxton	151	45.7–	57.3–	29.7—	52.6–
Ramsdale	173	75.8	82.8	62.1	67.1
Bellevue	164	85.2	79.2	79.8	70.6
Burton Joyce	300	129.4++	187.5++	170.5++	273.9++

+/– means p < .05; ++/— means p < .01

Source: Giggs et al. (1988)

ages up to the age of 25 years. They looked at the observed, and expected, number of pairs of patients who were resident in the same area and whose years of birth were within one year of each other. Results indicated that, until the age of about 15 years, there was little evidence of significant *space-time clustering* (box 6.3, page 184); however, between the ages of 16 and 20 (and for 18-year-olds in particular) there was very marked evidence of clustering. In other words, MS patients of a similar age lived close together in late adolescence to a much

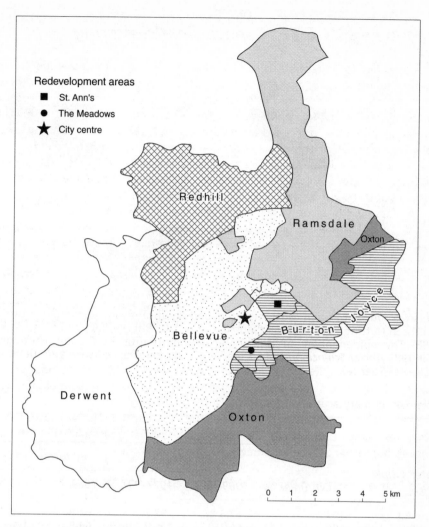

Figure 6.6 Water supply areas in Nottingham (Source: Giggs, J.A., Bourke, J.B., and Katschinski, B. (1988) The epidemiology of primary acute pancreatitis in Greater Nottingham: 1969–1983, *Social Science and Medicine*, 26, 79–89, reproduced with kind permission of Elsevier Science)

greater degree than chance would dictate. The tentative explanation is that MS is a delayed response (since patients will tend to present first in their thirties or forties) to a viral infection acquired, possibly via the exchange of saliva, in late teenage years. Merely mapping current place of residence would have failed completely to reveal this finding; a detailed knowledge of migration histories is required.

A third example comes from work by Sabel et al. (2000) on spatial variation in the incidence of motor neurone disease, or MND (also known as amyotrophic lateral sclerosis, or ALS) in Finland. MND is a progressive and ultimately fatal neurodegenerative disease, whose aetiology is unknown. It is a disease of low

BOX 6.3   *Space-time clustering*

We considered in Chapter 3 how to detect "clustering" of disease and illness using the addresses of those diagnosed. Often epidemiologists wish to know whether cases that are close in space are also close in time; we may need to know if there is *space-time interaction*. This is particularly important if we wish to establish whether or not a disease is infectious. If cases that live nearby were also diagnosed at, or about, the same time this would strengthen the argument for an infectious aetiology. Where there is debate about disease causation (for example, in leukaemia and lymphoma, or multiple sclerosis) some researchers have used statistical methods to detect such clustering or interaction.

A simple test is to form a table with two rows and two columns, the rows of which refer to whether or not a pair of cases are "close," and the columns denote whether or not a pair of cases are temporally "close." We then count how many pairs are close in time and space, close in space but not in time, close in time but not space, and not close in either. Statistically, we then determine whether the number of pairs close in time and space is significantly greater than expected on a chance basis. This is known as a *Knox test*. One difficulty is that a distance and time threshold must be fixed in advance. Are those living within five km (3 miles) "close" pairs, or should this be 10 km (6 miles)? Are those diagnosed within two days "close" in time, or should this be five days? There is some arbitrariness here, as well as dependence on the scale of investigation. In practise, researchers use a range of possible thresholds, or turn to other methods for detecting space-time interaction.

The research described in the text, on multiple sclerosis in Norway, takes these ideas one step further, since the authors look not at address at time of diagnosis but rather at where people lived at different ages.

See Thomas (1992) and Bailey and Gatrell (1995: 122–5) for further details.

incidence (approximately 1–2 per 100,000) and affects primarily older age groups. Data were collected on 1,000 deaths from MND between 1985 and 1995, matched by age and sex to population controls. Since the Finnish population register identifies all changes of address it was possible to look at where both cases and controls had lived since the mid-1960s. Using kernel estimation (see Chapter 3, pp. 62–4) it was possible to construct a relative risk surface according to current place of residence, and also former place of residence. It appears that those subsequently diagnosed with MND have, relative to people unaffected by the disease, spent many years living in the south-east of Finland, as well as inland from the city of Oulu and on the west coast, south of Vaasa. These are predominantly rural areas. Figure 6.7 illustrates this in a three-dimensional surface representation of the country, with light-shaded "peaks" denoting areas of increased risk, and darker "troughs" denoting areas of lower risk. Further, detailed epidemiological investigation is needed to shed light on the areas of elevated relative risk.

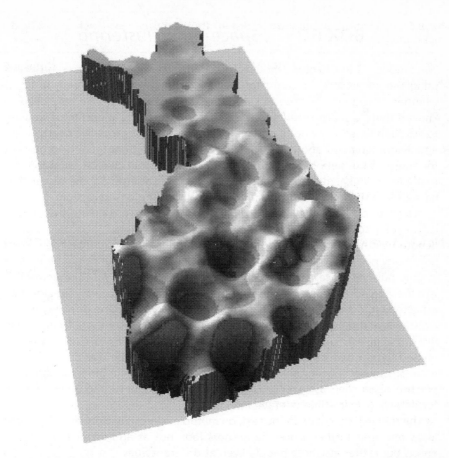

Figure 6.7 Relative risk of motor neurone disease in Finland (courtesy of Clive Sabel)

Impact of Health Status on Migration

I want here to consider other links between migration and health. Rather than looking at how migration affects health, I want to examine how health status affects the propensity to move. I want to consider the "health selection" hypothesis in more detail, and to ask what evidence there is that only the healthy may migrate. I also review evidence that suggests that people may migrate in order to gain access to health care or social support.

The Selectivity of Migration

Bentham (1988) considers the health selection hypothesis in some detail. He notes that migration rates tend to be higher for young, wealthy people employed in non-manual occupations; these people are likely to be healthier than the

Table 6.5 Percentage population in Britain that is permanently sick or disabled, by origin of migration

Age group	All residents	Within-district	Between-district	Between-region	Outside GB
21–24	0.7	0.3	0.3	0.2	0.1
25–29	0.8	0.4	0.2	0.3	0.1
30–34	0.9	0.7	0.5	0.5	0.2
55–59	6.3	11.1	6.6	8.8	2.8
60–64	11.2	18.2	11.9	12.0	4.8

Source: Bentham (1988), based on analysis of 1981 census data

Table 6.6 Stroke mortality in the United States: migrants and non-migrants (1979–81)

Race	Sex	US-born	US-born interregional migrants	Immigrants
White	Both	73.1	71.6	65.0
	Male	79.8	75.4	68.5
	Female	68.3	68.3	62.1
Black	Both	108.4	95.9	51.4
	Male	121.4	103.3	55.2
	Female	98.8	90.1	48.6

Rates are deaths per 100,000 persons per year
Source: Lanska (1997)

population at large, so that areas which are losing population tend to be those with higher mortality and morbidity. In the absence of information on migration, therefore, we do not know whether associations between ill-health and deprivation may be due in part to the selective loss of healthier people.

Bentham examines British data from the 1981 census on permanent sickness and disability and relates levels of morbidity to migration (where address has changed during the previous year). His results (table 6.5) suggest that, regardless of age, those moving the furthest (notably from overseas) have lower morbidity than the population as a whole; it is the "fitter" people who are moving. This is the "healthy migrant" effect. Among older adults, internal migrants tend to have higher levels of morbidity than the older population as a whole. Those moving shorter distances (within districts) have particularly high morbidity, suggesting that ill-health may be a possible factor in migration. This is evidence of "reverse" selectivity, in that it is the unhealthy rather than the healthy who appear to be moving.

What other evidence can be marshaled in support of health selection? Lanska (1997) showed that stroke mortality was significantly lower among immigrants (both white and non-white) to the United States than among the US-born resident population. Table 6.6 reveals that age-adjusted rates of stroke mortality are lowest among immigrants, but that those moving from one region to another also have lower rates than those who had not moved to another region. He argues that the costs, both financial and personal, involved in migration, operate to ensure that only the relatively healthy are "selected" for migration. No data are available, however, on the age at which migration occurred, or how many moves have been involved.

Other research in the United States, on maternal health, indicates that infant mortality is lower among the children of Puerto Rican-born and Hispanic Caribbean women than among those of Hispanic women born on the US mainland. These immigrant women are less likely to smoke or drink alcohol during pregnancy, though evidence indicates that these healthy behaviors disappear after arrival in the United States (Thiel de Bocanegra and Gany, 1997: 34).

Kington et al. (1998) looked at the "functional status" (impairment or health problem that limits daily activity) of people aged 60 years and over, who had moved from the southern states of the USA. The health of blacks who were born in the South but lived elsewhere was better than that of those born in the South who had not migrated. There was no difference in health status between Southern-born whites who migrated and those who stayed. The contrasting fortunes of the black Americans could not be explained by differences in socio-economic status. Kington and colleagues put this down to selective migration: the "healthy migrant" effect. The findings contrast with studies of mortality, which show that African Americans born in the South and migrating elsewhere, as well as those remaining in the South, have higher mortality rates than those born outside the South.

The selection hypothesis has been of particular relevance to studies of mental ill-health, especially in urban areas (Jones and Moon, 1987, Chapter 5). It has been argued that living in some inner city areas "breeds" or causes mental health problems, while others contend instead that the mentally ill migrate or "drift" to such areas. This latter hypothesis is one of selective migration. Loffler and Hafner (1999) study people suffering from schizophrenia in the German cities of Mannheim and Heidelberg. In both, there have been concentrations of schizophrenics in low-status inner-city areas for well over 25 years. But a close examination of individual biographies reveals that schizophrenics have suffered downward mobility into poor areas as the illness has disrupted their lives. Selective migration is therefore confirmed here.

Migration for Health Care and Social Support

This is a somewhat neglected area of research. The emergence of HIV and AIDS has produced some findings, while the extent to which older people retire in order to seek out social support and health care has been given some attention. We consider each briefly.

Ellis and Muschkin (1996) have examined the migration patterns of people with AIDS, looking specifically at moves from different states in the USA, to

Florida. Their particular interest centers on whether such people are moving to be close to informal sources of support, in particular elderly parents. As of late 1989 there were 9,555 cases of AIDS diagnosed in Florida, of whom 535 had moved from outside the state. While 28 percent of these migrants had moved to Miami, and others to urban areas, 20 percent had moved to small town and rural areas. Related work (Graham et al., 1995) points to high levels of in-migration to rural areas by HIV/AIDS sufferers. Statistical analysis of the Florida data shows that, even after adjusting for the effects of population size and exist-ing AIDS cases (since people with AIDS are more likely to move to areas where there are existing gay communities), the proportion of the population that is elderly has a significant influence on migration to particular counties. Although there is no direct evidence, "the most likely alternative motivation is to seek the care and support of family, primarily parents of post-retirement age" (Ellis and Muschkin, 1996: 1113). However, there are no data on individuals and so their conclusions must inevitably be rather cautious. Their own data suggest there is only a weak association between the concentration of an elderly population and the proportion of people with AIDS moving to particular counties ($r = 0.13$).

A more recent study (Wood et al., 2000) of people with HIV in British Colum-bia, Canada, shows that their migration to Vancouver is largely to be explained by their search for health care, although the culture of the city, including the comparative absence of prejudice there compared with other parts of the province, is also an explanatory factor. As in the Florida study, urban-rural migration was also accounted for by the need for social and family support. But in both studies the quantitative nature of the research means that explanation remains somewhat speculative. This is a research area that demands a qualita-tive perspective.

A good deal is known about patterns of migration among older people (Rogers, 1992), though detailed empirical work on the reasons for such moves is still required. Migration of the elderly is generally divided into that which is voluntary and that which is more constrained by personal circumstances, includ-ing health. In the first category are those who are recently retired, generally well-off and healthy, who seek a move for reasons of "amenity;" the classic example is the stream of migrants to Florida. The second type of older migrants com-prises those less able to care for themselves, or who anticipate a need for care. This group includes those moving to be nearer to children and other relatives, perhaps when losing a partner, as well as those who develop a disability or chronic illness that necessitates institutional care. Warnes and colleagues (1999) have looked at Britons retiring to southern Europe (Tuscany in Italy, Malta, the Algarve in Portugal, and the Spanish Costa del Sol) and reveal that the onset of chronic illness and disability was likely to encourage a return to Britain. However, this varied from place to place; for those on Malta, with relatively well-developed primary care services and a predominantly English-speaking population, a return due to illness was less likely than from other areas. In general, though, the expatriates painted a very positive picture of retiring to southern Europe.

The Relationship between Migration and
the Delivery of Health Services

Here, I want to consider the association between migration and the provision of health care to those moving from one place to another, usually from one country to another. To what extent do migrants get a "raw deal" from health care services?

Reijneveld (1998) looked at first-generation immigrants in Amsterdam, showing that those from Turkey, Morocco, and the former Dutch colonies made more demands on health care than non-immigrants. This reflected their poorer health status and was not to be explained simply in terms of lower income or occupational status. Other immigrant groups seem to make less use of preventive services, whether for screening or for ante-natal care, and where they do the communication difficulties between primary health care providers and immigrants act as a barrier to effective health care. Part of the reason for non-use of services lies in the illegal status of some immigrants and their fear of losing employment (for example, Moroccan agricultural and construction workers in Spain: Ugalde, 1997). Culturally sensitive programmes designed to address the health needs of migrants are likely to pay dividends. For example, Verrept and Louckx (1997) demonstrate that health "advocates," Moroccan and Turkish women recruited and trained as health workers, have improved the health status of immigrants to Belgium from Morocco and Turkey.

Access to health care in the USA among the foreign-born is limited by their lack of health insurance (Thamer et al., 1997). In 1990 the foreign-born population was twice as likely as the US-born population to be uninsured (26 percent compared with 13 percent), with those of Hispanic origin having an uninsured rate of 41 percent. However, in some parts of the US, these figures are markedly worse. For example, in New York City survey research (Sun et al., 1998) suggests the proportion of uninsured foreign-born people may be as high as 77 percent. The lack of a regular income and health insurance, and therefore inability to pay, means that immunization uptake is low (only 46 percent among the foreign-born). But efforts are being made by public health authorities to "target" such vulnerable immigrant populations, with immunization being offered free to children of immigrant families in New York City (Sun et al., 1998). In addition, the New York Task Force on Immigrant Health (NYTFIH) brings together academics, community activists, and health care providers to increase access to culturally sensitive health care (Thiel de Bocanegra and Gany, 1997). The NYTFIH has developed outreach strategies to make contact with immigrants likely to be suffering from tuberculosis, for instance by approaching them through community groups or on the street (where many will be street vendors).

In other settings explanations for a low uptake of health services have little to do with material resources. For example, women moving to Israel from the former Soviet Union seem to avoid making use of preventive services, such as screening for breast or cervical cancer, even though they were using such services before they moved (Remennick, 1999). This may reflect difficulties in adapting to a new culture and environment, as well as linguistic and cultural barriers to accessing services. The pattern is also observed among those from Vietnam now living in London (Free et al., 1999). Qualitative (focus group) research was used to study community groups and this revealed that a lack of knowledge of out-of-hours arrangements in general practice, arising from communication difficulties, meant poor access to primary health care. Research on Filipino immigrants to the USA (Yamada et al., 1999), also using focus groups, suggests that the cost constraints in getting help for the treatment of tuberculosis are less serious than the wish to deny or hide the disease or to put faith in traditional treatments. Thus, linguistic barriers, cultural differences, and relative social and geographical isolation all serve to reduce the ability of migrants to gain access to health care services.

Other qualitative work has explored issues of gender and "race" in a more sophisticated way, drawing on feminism in particular to shed light on the health-seeking behaviors of first-generation women immigrants to Vancouver (Dyck, 1995b). The women studied by Dyck are from Hong Kong and the Chinese mainland, as well as from India and Fiji. But Dyck cautions against the simple coding of "race" or ethnicity as an explanatory variable, arguing that it is a social construct. The attachment of simple labels, she argues, "objectifies" these groups and serves both to stereotype individuals as passive victims of cultural change (rather than people with identity and human agency) and to "deflect attention away from the contributions of political and ideological processes, including the power relations of health service provision" (Dyck, 1995b: 249). Dyck considers in detail the case of one woman who seeks help from a variety of sources of health care, both traditional Chinese healers and a Chinese-Canadian family doctor (physician), but who, like others, is able to access informal social networks for health care. Such health care may be accessed locally, but some women draw on help and advice from those still living in China and the Indian sub-continent. Clearly, health care may be sought from a variety of "places."

It is worth noting that the migration of health professionals themselves has a potential impact on the availability of health care. In Britain, there are long-standing concerns about the retention of a qualified medical (and nursing) workforce, and particularly about how movements of health professionals might cause inequalities in the supply of health care. In some areas staff turnover is considerably higher than elsewhere, and this is associated with material deprivation (Taylor and Leese, 1998). However, this turnover seems to be due less to the movements of general practitioners from one region to another and more to their simply leaving the profession. Furthermore, drawing on qualified staff from overseas, particularly the developing world, simply denudes such countries of their own much-needed health professionals.

Concluding Remarks

We have considered a number of different themes in this chapter, concerning the relationship between migration and health status, and between migration and health care delivery. It is quite clear that while there is a considerable literature on the former, research on the associations between migration and service provision has been much less substantial. What conclusions can we draw from the material presented here?

I think it is fair to say that much of the work reviewed here draws on essentially positivist approaches, creating statistical models of social epidemiology in which the "risk" of migration is examined after attempting to adjust for other factors. Clearly, it is important to do this, otherwise we will obtain bogus, or at least counter-intuitive conclusions. For example, Wei et al. (1996) present evidence that Mexican-Americans born in the United States have higher mortality rates than those born in Mexico. But this surprising finding is due largely to socio-economic confounding, since members of the latter group were of much lower socio-economic status. Once adjustment is made for this, the differences lessen substantially.

There is clearly scope for alternative approaches to migration and health. We have touched upon some of these here. For example, an understanding of the "experience" of migration and the health impacts of the event itself requires qualitative methods that may be located within a social interactionist or structurationist framework. And studying the health impacts of labor migration becomes impossible without engaging with a structuralist approach that looks at the political economy of health.

Regardless of the explanatory framework we adopt, many of the studies considered here have been cross-sectional, perhaps studying samples of migrants and non-migrants at one point in time. We saw some benefits to be gained by studying the health status of people before and after their moves (illustrated by the Tokelau Islanders), but this kind of longitudinal study is relatively rare. It is certainly required in order to resolve debates about health "selection."

Health and migration are intimately linked. Given that migration is an inherently social and geographical process, and that health and health care are socially and geographically patterned, this is hardly surprising. Yet much more work needs to be done to clarify the relationships, and in particular to flesh out some of the ways in which migration impacts upon, and is affected by, health care delivery.

FURTHER READING

A good, contemporary introduction to geographical research on migration is provided in Boyle et al. (1998). This book also considers in more detail some of the types of migration touched on above, such as forced migration. On this subject, see the impassioned essay by Arundhati Roy (1999) on the human dislocations produced by dam construction in western India. See also King (1995).

Gellert (1993) offers a good overview of the impact of international migration on the spread and control of communicable diseases. The chapters by Little and Baker, and by Kaplan, in the book edited by Mascie-Taylor and Lasker (1988), are well worth reading. There is a considerable literature on the differences in mortality among migrant groups. For a comprehensive picture for England and Wales, see Harding and Maxwell (1997). For a review of research on migration and health in Europe see Carballo et al. (1998).

PART III

HEALTH AND HUMAN MODIFICATION OF THE ENVIRONMENT

CHAPTER 7

AIR QUALITY AND HEALTH

The main aim in this chapter is to review some of the associations between air quality and ill-health. In speaking of air "quality" I am, of course, speaking mostly about air "pollution;" but reference to air quality permits me to discuss some of the interesting work on radon, a natural radioactive gas. Much of the literature in this area tends to adopt a classical positivist epidemiological stance. However, we shall see that some researchers call for a better understanding of "lay beliefs" about such links, while a strong case can be made for a structuralist interpretation of the most serious pollution occasioned by industrial "accidents."

I have chosen to organize the discussion in terms of the *source* of pollution. Some pollutants, such as radon gas and low-level ozone, are rather diffuse in origin; they may be widespread and can be considered therefore as "areal" sources. Others, such as the pollutants accompanying vehicle exhaust emissions, arise from "linear" sources; principally the roads on which traffic flows. I shall thus consider the evidence linking proximity to major thoroughfares to ill-health. Last, I shall review a selection of research studies that examine "point" sources of industrial pollution; to what extent are those living near incinerators, coking plants, power stations, and other industrial sites more at risk of disease and ill-health than those living elsewhere? As with any classification, this three-fold division is somewhat simplistic; for example, vehicular emissions diffuse over a wide area and are hardly confined to the roads themselves! Nonetheless, it serves as a convenient organizing framework.

A variety of health effects will figure here. Of particular interest is respiratory disease such as asthma, since plenty of evidence exists that the incidence of asthma has increased over the past 30 years. Data for the UK (table 7.1) show that the rate of consultations for asthma have increased four-fold since the early 1970s, for both males and females. Some caution needs to be exercised, however, since it may be the case that "asthma" is becoming more commonly used as a label for respiratory conditions that were previously recorded otherwise. Whether this is able to account for the fact that prescriptions for asthma are 50 percent higher in north and north-west England than in other parts of the country is debatable.

Table 7.1 Increase in asthma consultations in England and Wales

	1971–2	1981–2	1991–2
Males	10.6	20.0	42.9
Females	8.6	15.9	42.2

Rate of patient consultations per 1,000 persons

Source: *Department of Health Epidemiological Overview on Asthma* (1995)

Other forms of air pollution have been implicated in lung cancer and in cardiovascular disease, the hypothesis being that the pollutants enter the blood stream from the lungs and are carried to the heart; some heart attacks may, therefore, be triggered by air pollution. From a geographical point of view we are particularly interested here in studies that ask whether where you live affects exposure to, and health damage from, poor air quality. But other important studies adopt a time series approach, looking at temporal data on air pollution and corresponding patterns of morbidity and mortality.

Types of Pollutants

Early work on air pollution and health in the developed world focused on sulphur dioxide (SO_2) and smoke, since these were the main pollutants in urban areas before the 1970s. The principal source of SO_2 is the burning of fossil fuels, but the reduction in burning coal for domestic use has reduced emissions over the past 30 years. However, coal-fired power stations continue to be a major source of the pollutant. More recently, and with the growth of motor vehicle traffic, concern has shifted to pollutants such as nitrogen oxides (NOx) and volatile organic compounds (VOCs). The principal source of NOx is from motor vehicles, though power stations and other industrial processes contribute to the load. VOCs include benzene, emitted from petrol fumes and vehicle exhausts. The presence of sunlight causes NOx and hydrocarbons (the residues of partly burnt fuel) to combine with other gases in the atmosphere to form ozone (O_3), which we consider below.

Particulate air pollution refers to a mix of solid particles and liquid droplets. In contemporary society, motor vehicle exhaust emissions (especially from diesel engines) are a significant source. Interest centres on particles that can be inhaled (with a diameter less than 10 μm: so-called PM_{10} particles), and especially on the finer particles (diameter less than 2.5 μm: $PM_{2.5}$) that can be breathed deep into the lungs.

In the present chapter, as well as the following two chapters, we are looking at aspects of the field known as *environmental epidemiology*; this deals with the relationships between environmental quality and health outcomes. In this

BOX 7.1 *Exposure assessment*

A simple way of estimating exposure is simply to record whether someone lives in the same area as the source of air pollution. For example, if a county contained a suspected point source of pollution we could suggest that all those living in the county were potentially exposed to pollution. But this is clearly rather crude, since the movement of polluted air will not respect administrative boundaries! We could look at proximity to pollution sources, and distance is often used as a surrogate measure of exposure; as we shall see, some work looks at proximity to main roads, or to point sources of pollution. Better still would be to identify areas of risk using an air dispersion model, since the pollutant will not disperse uniformly in all directions. Or, rather than modeling air pollution, we could monitor levels of air quality at a set of locations. Issues then concern where to locate monitoring sites, and how many to use. Finally, in an ideal world, we would measure directly the exposure of a set of individuals, using personal monitors; sophisticated (and well-resourced) studies do so. There are different ways of characterizing "exposure," but the more effectively we try to do this the more costly it becomes.

Local meteorological conditions can affect the dispersion and subsequent concentration of outdoor pollutants, while human spatial behavior over the course of the day and week will serve to affect personal exposure and subsequent "dose." Our primary concern here is what the health effects of such doses might be. While laboratory work can assess the possible effects of single pollutants, and provide controls for a host of potential confounding variables, "air pollution" rarely consists of exposure to a single chemical; rather, we are invariably exposed to a mix of pollutants.

chapter we are examining some of the evidence concerning the impact of air quality on human health (particularly respiratory disease and cancers). A key issue here, as in all environmental epidemiology, is to characterize the nature of the exposure. Levels of air quality need to be measured, or estimated. We need, ideally, to look at how a pollutant disperses from its source, and the extent to which people come into contact with it. How is the pollution distributed in time and space? These issues are considered briefly in box 7.1.

I do not consider indoor sources of air pollution here, though it should be appreciated that faulty gas fires and cooking appliances can be a major health risk, especially if poor ventilation traps the gases. Nitrogen dioxide is a common indoor air pollutant and exposure to 400 ppb (parts per billion), as is common when cooking on a gas stove, can exacerbate the adverse reactions of mild asthmatics who are already allergic to house dust mite (Tunnicliffe et al., 1994).

Area Sources

I begin by looking at one important American study, and then focus on two specific sources of air pollution. The former is research conducted by Dockery and his colleagues (1993), examining the association between mortality and air pollution in six American cities. It was a prospective study, following adults from the mid-1970s until death (where this had occurred). Access to individual-level data allowed the researchers to examine the effects of air pollution on mortality, while adjusting for the effects of age, sex, occupation, and, most crucially, smoking status. The six communities were: Watertown (Massachusetts); Steubenville (Ohio); Portage (Wisconsin); Topeka (Kansas); Harriman (Tennessee); and specific parts of St Louis (Missouri). Over 8,000 individuals were recruited to the study in 1974, of whom 1,430 had died by late 1989. Outdoor concentrations of air pollutants (total and fine particulates, SO_2, O_3, and sulphates) were measured at a central air-monitoring site in each community, at various dates.

Results (figure 7.1) show mortality rates (adjusted for age, sex, smoking, education, and body mass) plotted against average pollution levels in each city. The correlation with fine particles is particularly striking. Results are essentially unchanged when looking at specific causes, such as cardiovascular and respiratory disease. In policy terms, the message is quite clear: we need to reduce urban air pollution.

Radon

Radon is a radioactive gas formed during the decay process of uranium-238 and thorium-232, naturally occurring nuclides found in minerals in the earth's crust. Radon decays to produce further isotopes (known as radon "progeny"), which may attach themselves to water or to aerosols such as dusts. Upon reaching the open air the gas mixes with the atmosphere, but if trapped in underground mines or in houses, concentrations can build up. Since uranium is found in some igneous rocks (especially granites) and also some sedimentary rocks, radon levels are, on a broad scale, very variable from place to place. Within Britain, high concentrations (measured in Becquerels per cubic metre, Bq/m^3) are found in much of Cornwall, parts of west Devon and west Derbyshire (near Buxton), central Northamptonshire, and central Grampian in Scotland (Kendall et al., 1994). In some of these areas 30 percent or more of the housing stock has radon levels in excess of $200 Bq/m^3$ (the "action" level set by the National Radiological Protection Board in the UK, averaged over the year). In the USA (where the action level is $148 Bq/m^3$), high levels of radon are found in some counties within Iowa, North Dakota, New Mexico, Minnesota, Nebraska, and Pennsylvania. Levels are elevated especially where the soil or rock is permeable and the gas can move through to the surface. At much more local scales radon levels are highly variable, depending on building construction and ventilation. Where a building has cracks or fissures the gas finds a route of entry and may be trapped

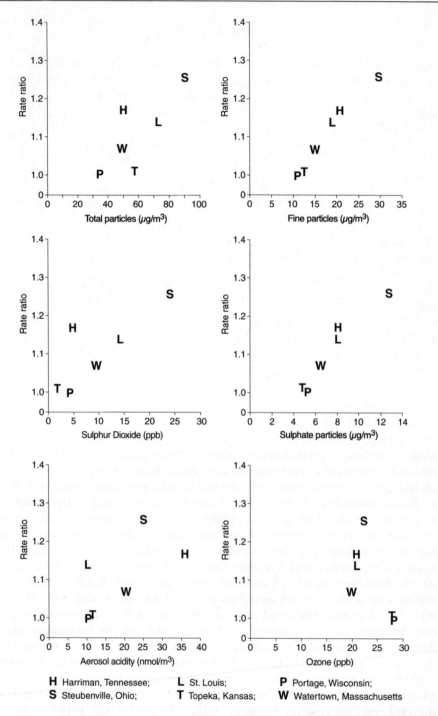

Figure 7.1 Relationship between estimated adjusted mortality ratios and pollution levels in six US cities (Source: Dockery, D.W., Pope, C.A., Xu, X., Spengler, J.D., Ware, J.H., Fay, M.E., Ferris, B.G., and Speizer, F.E. (1993) An association between air pollution and mortality in six U.S. cities, *New England Journal of Medicine*, 329, 1753–9, reproduced with kind permission of the Massachusetts Medical Society)

if there is a low air exchange rate and the building is well sealed. Thus radon concentrations may vary between adjacent dwellings. Levels also vary temporally, both diurnally (peaking in the early morning) and also seasonally (being elevated in winter months).

Very high levels are recorded in mines and in caves. For example, the average level in a sample of British caves was $2,900 \, Bq/m^3$, while the Giant's Hole cave in Derbyshire recorded $46,000 \, Bq/m^3$ over the course of a year. Indeed, it was studies of underground miners of uranium and other ores that generated models of the relationships between radon exposure and lung cancer. Pooling a series of 12 studies of miners results in an observed total of over 2,600 lung cancers in a cohort of 60,000 miners, where only 750 cancers were expected (Kendall et al., 1994: 4). These studies of miners have been used to extrapolate their very high exposures to much lower indoor ones and to suggest, in the US at least, that between 6,000 and 36,000 lung cancer deaths each year (the US Environmental Protection Agency suggests 15,000) are caused by residential radon exposure (Lubin et al., 1995). Despite early misgivings about such extrapolation the current consensus is that studies of underground miners remain the best source of data to use in assessing the risk from indoor radon (Lubin and Boice, 1997).

The primary danger of the gas lies in the radioactive decay process and the subsequent production of the radon progeny (such as polonium 214 and 218, bismuth 214, lead 214); the aerosols may lodge in lungs, exposing the lung tissue to alpha and gamma radiation. It is for this reason that the main health risk is considered to be lung cancer. Some studies have sought to demonstrate a link using aggregate data for sets of areal units, while other, more satisfactory, studies consider individual risk via case-control studies.

An example of an aggregate study is that by Haynes (1993), who looked at the areal association between lung cancer mortality and radon in England and Wales, with particular attention to the counties of Cornwall and Devon in southwest England. Using as areal units the 401 local authority districts in England and Wales, Haynes performs a regression analysis of lung cancer mortality (for the years 1979–83) on three covariates: an estimate of smoking prevalence; social class; and population density. All variables are highly significant. He then relates the residuals from this analysis (the unexplained variation) to average levels of radon in Cornwall and Devon districts, reasoning that, having adjusted for some possible risk factors for lung cancer (of which smoking is the most obvious), any districts where observed mortality exceeds that predicted by the model might be those where radon levels are high. However, there is little association between the residuals and average radon levels in the districts (figure 7.2); indeed, among women there is a suggestion of lower than expected lung cancer in districts with the higher levels of radon. Other aggregate geographical health studies suggest possible links between radon and other cancers, such as prostate cancer in men and leukaemias in children, and even between radon and motor neurone disease (Neilson et al., 1996) but these studies are hotly disputed. A study by Richardson et al. (1995) of 459 local authority districts in Britain finds no evidence of an association between childhood leukaemia and radon.

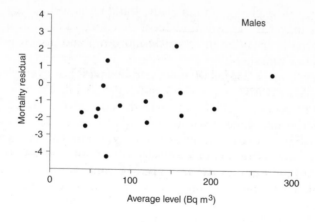

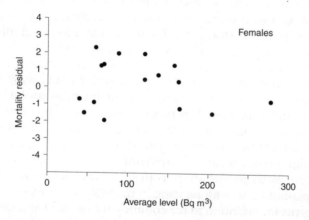

Figure 7.2 Relationship between lung cancer residuals and radon levels in Cornwall and Devon districts (Source: Haynes, R. (1993) Radon and lung cancer in Cornwall and Devon, *Environment and Planning A*, 25, 1361–6, reproduced with kind permission of Pion Ltd, London)

The problem with these studies is that they fail to provide adequate controls for confounding. As Goldsmith (1999) has shown, lung cancer rates for US counties are inevitably *negatively* correlated with radon levels measured for such counties. Residential radon levels are higher in low-density, suburban residences (where income is higher and people are willing to pay for measurements) and are therefore negatively associated with population density. Since population density is well known to correlate positively with lung cancer, it follows that aggregate levels of radon will necessarily be inversely correlated with lung cancer. Thus population density confounds the relationship between the independent variable of interest (radon) and the dependent variable (lung cancer). In addition, some aggregate studies are based on relatively few radon measurements. For example, the measure of exposure in the Richardson

Table 7.2 Relative risk of lung cancer in Stockholm women, by radon exposure

Age group	Cumulative radon exposure (Bq/m³/year)		
	1251–2500	2501–5000	More than 5000
All ages	1.4	1.7[a]	2.3[a]
Under 55 years	1.9	2.1	5.2[a]
55–64 years	1.3	2.1	2.3
Over 65 years	1.1	1.2	1.2

[a] Relative risk significantly greater than 1.0 (95 percent level)

Source: Pershagen et al. (1992)

(1995) study used less than ten measurements in each local authority area. Since we know that radon levels are highly variable over short distances this seems very questionable and I suggest that there is much more to be gained from detailed, individual-level, case-control studies. The advantage of these is that they can adjust for smoking, the obvious and well-established risk factor for lung cancer.

Among these, Pershagen et al. (1992) has conducted a study of women living in Stockholm, Sweden, a country where the radon problem has received considerable attention because of the geological structure. The authors studied 210 women diagnosed with lung cancer between 1983 and 1986. Age-matched controls were taken from both hospitals and the community (191 and 209 women respectively). All women were interviewed in order to obtain data on smoking behavior and previous residences (for all of which radon measurements were obtained). The results show an increase in the risk of developing lung cancer as residential exposure to radon increases, even when making allowances in the figures for smoking. For example, there is a relative risk of 2.3 (with a 95 percent confidence interval of 1.1–4.5) for women with a lifetime cumulative exposure of >5,000 Bq/m³/yr (table 7.2); regardless of age, the increase from low to high radon exposure generates a consistent increase in the risk of lung cancer. Among heavy smokers (more than ten cigarettes per day) the relative risk is 13.6 for those exposed to <2,500 Bq/m³/yr but 21.7 for those exposed to twice the radiation. It seems clear that residential radon exposure affects the risk of developing lung cancer. It further seems clear that the problem will remain for many years, since more recently built properties have low ventilation rates in order to save energy.

The advantage of working in a country such as Sweden is the availability of detailed records of residential histories. This allows us to look at likely cumulative exposure. Work in other Scandinavian countries such as Finland (Teppo, 1998) confirms that there is an elevated risk of lung cancer after exposure to radon, though work in south-west England (Darby et al., 1998), also looking at radon concentrations in the previous homes of cases and controls, indicates that the effect is quite modest. As we saw in Chapter 6, we need data on migration histories to gain accurate measures of possible exposure.

Table 7.3 Estimate of population exposure (in millions) to ozone episodes in Britain (1987–90)

Average hours per year >90 ppb	Rural areas	Urban areas	Total
4	20.8	24.4	45.1
8	20.2	24.4	44.6
12	19.6	23.8	43.4
16	18.0	21.3	39.3
20	8.1	7.8	16.0
24	1.8	0.4	2.2

Source: UK Photochemical Oxidants Review Group (1993: 89)

Ozone

Ozone (O_3) is a summer pollutant, at its worst in more rural areas and when temperatures are high and wind speeds are low. In Britain, concentrations tend to be higher in the south of England and in Wales. There is very substantial temporal (both daily and seasonal) variation, with ozone "episodes" in summer months; the maximum concentrations are in the afternoon, with lower levels in early morning. Air quality goals for O_3 vary from country to country; in the UK an hourly average of 90 ppb is regarded as "poor." In Australia the hourly air quality goal is 120 ppb, while in Japan it is 60 ppb. In California the figure is 90 ppb.

Some estimates of population exposure to high levels of O_3 have been made. In the UK, research (UK Photochemical Oxidants Review Group, 1993) indicates that large numbers of people may be exposed to more than 16 hours of high ozone concentrations during a year (table 7.3). There are marked urban-rural differences, with 1.8 million people in rural areas experiencing more than 24 hours above 90 ppb, compared with only 400,000 in urban areas.

Animal studies suggest that ozone damages the respiratory tract, although this damage is likely to be repaired after short-term exposure has passed. Controlled experiments on humans show that lung function (measured by forced expiratory volume, or FEV, the maximum volume of air that can be breathed out in one second) is impaired as duration and concentration of exposure increase. Vigorous exercise worsens the effect, since this increases the volume of air inhaled and allows the ozone to penetrate further. Studies of children attending summer camps in the USA have suggested that exposure to 100 ppb of O_3 leads to a reduction of about 3 percent FEV (UK Photochemical Oxidants Review Group, 1993) while adult studies suggest equivalent, or worse, effects on lung function.

Particular interest has centred on the possible link between asthma and ozone, given that the incidence of asthma has increased over recent years (table 7.1). When asthmatics are exposed to O_3 they show a greater response to allergens (such as pollen and house dust mite) than those exposed to filtered air; it is thus

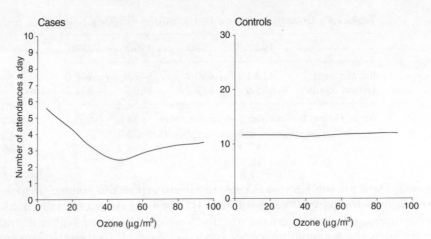

Figure 7.3 Relationship between number of children attending hospital Accident and Emergency services for acute wheeze (cases) and other reasons (controls), with ozone levels (Source: Buchdahl, R., Parker, A., Stebbings, T., and Babiker, A. (1996) Association between air pollution and acute childhood wheezy episodes: prospective epidemiological study, *British Medical Journal*, 312, 661–5, reproduced with kind permission of BMJ Publishing Group)

thought that O_3 increases sensitivity to pollen. Woodward and his colleagues (1995) report on a study in Houston, USA which showed that the probability of an asthma episode increased as ozone exposure increased. Ozone does not appear to "cause" asthma; what it does is to allow allergens (such as house dust mite, pollens, and animal proteins from cats) to penetrate the airway wall by damaging its lining. The airways are then sensitive to stimuli such as cold air, irritant fumes and smoke, and even exercise.

Anderson and his colleagues (1996) have examined associations between ozone levels and mortality (all causes, respiratory, and cardiovascular) in London between 1987 and 1992. They look at the increase in mortality associated with increases in maximum hourly ozone concentrations, over a range of about 2 to 45 ppb. Regardless of season there are striking increases in respiratory mortality of about 5–6 percent. They argue that ozone exposure hastens death in people who are already seriously ill. Other work by the same research group has shown that high ozone levels on a given day lead to increased rates of hospital admissions for respiratory disease on the following day. Research on children in London seeking urgent care for acute episodes of wheeze (Buchdahl et al., 1996) demonstrates a U-shaped relation between attendances and ozone levels. Compared with controls (figure 7.3) the number of cases increases after exposure of about 50 ppb (mean 24-hour concentration).

Linear Sources

Although pollution from industrial and domestic sources in the developed world has decreased over the last 30 years, that from mobile sources has undoubtedly

Table 7.4 Growth in vehicle traffic in USA (1970–96)

	1970	1980	1990	1996
Rural roads	424.1	450.7	547.9	540.0
Urban roads	495.6	671.2	869.9	937.7

Note: Figures are in billions of vehicle miles
Source: US Department of Transportation (1999)

increased. Data for the US reveal substantial growth in the volume of passenger traffic over the past 40 years (table 7.4). The increase is especially marked in urban areas, where traffic has virtually doubled. Given the higher densities in urban areas the potential for heavy exposure of people to the emissions from such vehicles has clearly risen dramatically.

I want here to examine the evidence that links vehicle exhaust emissions to health problems. Such exhaust emissions include carbon monoxide, nitrogen oxides, and unburned hydrocarbons (all "primary" pollutants). We have already noted that the presence of sunlight causes the hydrocarbons and oxides of nitrogen to produce ozone, a "secondary" pollutant. In addition, while diesel engines emit no lead, and much less carbon monoxide than petrol engines, they emit small particulates.

Despite the considerable level of public interest in the possible links between exhaust emissions and ill-health there have been relatively few empirical studies. Partly, this is because of the severe problems involved in capturing exposure adequately.

One of the classic studies was done by a group of German epidemiologists (Wjst et al., 1993) who looked at the association between road traffic in Munich and its impact on the lung function and respiratory symptoms of children aged between 9 and 11 years. A questionnaire sought information on whether asthma had ever been diagnosed and on recent symptoms such as "runny nose" and "frequent cough," together with demographic data. Data were also available on parental history of asthma, smoking in the home, use of coal or gas in the home and other potential confounders. Of the 4,678 children, various measures of lung function were also collected. Data on exposure was much weaker and was assessed by use of maximum daily traffic counts in 117 school districts; the exposure of any child was therefore taken to be that of the street with the highest volume of traffic, in the school district of residence. Traffic counts ranged from 7,000 to 125,000 vehicles during a 24-hour period. Of the lung function measurements, peak expiratory flow rates were reduced by 0.71 percent for every increase of 25,000 cars per district (with a confidence interval of 0.33 to 1.08 percent), after adjusting for confounders. For respiratory symptoms there was no significant increase in asthma diagnosis, but a small (though significant) increase in minor respiratory infections.

In common with most other studies, this work is cross-sectional rather than longitudinal, and the measure of exposure, which assumes the same exposure for all children living in a district, is undoubtedly crude. The same criticism can be made of a British study (Whitelegg et al., 1995) although this improved slightly on the measure of exposure. The authors adopted a hierarchical sampling scheme, selecting for study ten local authority districts in northern England and Scotland. Within each, roads were selected for which traffic count data were available and where there was a range of traffic volumes (from about only 200 vehicles per day along small streets, to over 65,000 along the busiest thoroughfares). Site visits ensured that the 57 streets chosen were residential, that there was a 30–40 m.p.h. speed limit on each, and that there were no other obvious sources of pollution in the vicinity. Health data were obtained from a household survey, using a self-completed symptom checklist; this asked about the presence of common symptoms, some of which (such as sore throat or cough) one might expect to see associated with air pollution, others (such as cuts and grazing) which one would not. Data were also collected on possible confounders, such as smoking and household damp. Although the response rate was low (38 percent of the 1,916 houses that were mailed with the questionnaire responded) there were some intriguing findings resulting from a logistic regression analysis. For symptoms such as "blocked or runny nose," "sore or red eyes," and "cough" there was a significant effect of traffic level, adjusting for household and personal characteristics, though for "breathing difficulty" there was not. For example, the risk of "runny nose" was 1.5 times greater on a road with 50,000 vehicles per day, compared with a road with only 5,000 vehicles.

Others have adopted different measures of exposure, as well as different health outcomes. For example, Brunekreef and his colleagues (1997) looked at lung function among Dutch children attending schools near motorways. PM_{10} and black smoke levels were measured in the classrooms. There were reductions of up to 8 percent in lung function among children attending schools with the highest pollution levels. Wilkinson and his colleagues (1999) conducted research in north-west London, taking as the health outcome hospital admissions for asthma among children aged 5–14 years. These cases were matched with controls – children admitted to hospital for other, non-respiratory, emergencies. The risk of admission for asthma was assessed in relation to three proxy exposure variables: distance from the postcoded residence to the nearest main road; distance to the nearest main road that had (estimated) peak hour traffic flows of more than 1,000 vehicles; and the estimated traffic volume along all roads within 150 metres of the postcoded residence. Adjustment (control) was made for deprivation in the enumeration district in which the child lived. For none of the exposure variables were the odds ratios significantly different from unity, suggesting that there is no association between risk of admission for asthma and traffic-related pollution. However, the exposure measures are somewhat crude and, in common with most other studies, take no account of exposure indoors (the Whitelegg study made an attempt to do so) nor of the activity spaces of the children. Adjustment for the confounding effect of smoking was done simply by using a deprivation score that is taken as a correlate of smoking. Most seriously, the health outcome measure represents only the "tip of the iceberg" and although dealing with more than 2,000 children neglects the vast majority who suffer from asthma but manage the condition themselves or with the help of primary care professionals.

Some writers have speculated about a possible causal link between exposure to benzene and the incidence of leukaemia, and geographical studies on this are emerging. One, in the West Midlands region of England, suggests that childhood leukaemia is elevated close to main roads and petrol (gasoline) stations, though the raised risk is not statistically significant (Harrison et al., 1999).

While environmental epidemiologists have begun to study the way exhaust emissions affect those living near roads, other researchers have attempted to assess the exposure of road users to air pollutants. Concentrations of nitrogen oxides, carbon monoxide and VOCs inhaled by car users may be two or three times higher than those inhaled by pedestrians or cyclists. Benzene concentrations in vehicles are 18 times higher than background levels, and exposures among those people who use petrol (gasoline) pumps very frequently may be particularly high.

It should also be noted that since some air pollution is more common along very busy roads and road junctions, and these are usually places which are disadvantaged in other ways (poorer quality housing, for example) then the burden of pollution may fall disproportionately on poorer groups, including those who do not necessarily contribute to the problem. Issues of environmental health "equity" arise here, as they have in other areas considered in this book.

Point Sources

Here I assess the health risk of living near possible point sources of air pollution. Such point sources may come in a variety of forms. They can include: incinerators, such as those used for domestic (municipal), industrial, hazardous, and nuclear wastes; landfill sites (though these may be more a source of water pollution than of air pollution); industrial and manufacturing plants such as smelters, cement works, and fertiliser and pesticide factories; and power stations, such as those fueled by nuclear materials. I shall consider a small set of empirical studies drawn from a larger set, with particular attention to those where the pollution results from an acute, sudden episode. I begin with an account of one such disaster.

Point Sources in the Developing World

During the night of Sunday December 2, 1984, there was a leak of methyl isocyanate and hydrogen cyanide gases from the Union Carbide pesticides plant on the outskirts of the city of Bhopal, in the central Indian state of Madya Pradesh. Two thousand people were dead the following morning, while about 500,000 people were caught by the gas cloud, many of whom have continued to suffer since from emphysema, asthma, tuberculosis, and eye disease. Of these,

300,000 have been awarded "compensation" (with a maximum payment of under $10,000), sourced from a fund of $470 million which Union Carbide agreed, five years after the event, to pay (Mukerjee, 1995).

Health impacts include imprints etched on the eyes of those running during the night to escape the gases. As the author of a report commissioned by the Indian Council of Medical Research, M.P. Dwivedi, stated, "as they ran, the gas hurt their eyes, so they shut them as tight as possible, just leaving a slit to see through. Today, that slit, an opaque line on their corneas, is permanent" (in the *Guardian* newspaper, 13 August 1998). No-one knew that, since methyl isocyanate reacts readily with water, a wet cloth placed over the face would have prevented the gas from penetrating the lung. Fifteen weeks after the disaster, researchers found that 38 percent of 260 people still living within two kilometers of the plant had "burning eyes," 19 percent had poor vision, and 6.5 percent had corneal opacity of the sort described by Dwivedi (Mukerjee, 1995). Three months afterwards, 39 percent of those living nearby had some form of respiratory impairment. Other work has compared pregnancy outcome among Bhopal women with those in an unexposed area (Bhandari et al., 1990). The rate of spontaneous abortions (miscarriages) in the exposed group was 24.2 percent, compared with only 5.6 percent in the control area. In a random sample of 454 adults, stratified by distance from the plant, Cullinan et al. (1997) found severe respiratory problems ten years after the event, with the frequency of symptoms declining with distance.

There have therefore been several classical epidemiological studies into the effects of the Bhopal disaster. However, one can argue that while this is of "scientific" interest it has had little impact on compensation claims and sheds no light on why the disaster happened. To do this, it is much more illuminating to reflect on the nature of pesticide production in India and wider issues of the political economy of development in the third world, issues discussed fully in Bogard (1989) and Shrivastava (1992); we might, in other words, gain a better understanding from the kind of structuralist perspective outlined in Chapter 2.

The license to import and use methyl isocyanate at the Union Carbide plant was based on its role in the manufacture of pesticides needed to increase food production as part of India's "Green Revolution." But we can also see the plant's location as being determined more by the need to transfer, wherever possible, hazardous production technologies from the developed world to developing countries. Increasing state regulation of such technology in the core leads to falling profits and pressure to relocate where cheap labor is available (the population of Bhopal rose dramatically after the Union Carbide plant was built). Yet those living near the plant knew little or nothing of its function or the potential hazard, while those employed there did not know how to handle non-routine events. "Not in my backyard" was not an option that the poor of Bhopal could afford, while the fact that the plant was relatively unimportant to the parent company meant that investment, particularly in safety and emergency planning, was very limited. Uncertainty was embedded in the fabric of the plant and the locality, and was used to tolerate risk. In addition the hazards associated with pesticide production were mitigated (compensated) by the need to lessen the impact of food shortages and unemployment. For one author at least, the world system of capitalism is to blame. "Political tradeoffs to enhance legitimacy and global economic demands generated by the capitalist imperative of accumulation narrowed the chances for detecting the dangers at Bhopal" (Bogard, 1989: 104). To speak of Bhopal as an "accident" neglects the role played by human agents operating thousands of miles away, or the culpability of the global chemical industry.

Point Sources in the Developed World

No industrial disaster in the developed world matches the impact on death and morbidity of the Bhopal explosion. The potential health risks of a major nuclear accident are, of course, a serious concern, and I begin this section by considering what, if any, were the health impacts of an accident at a nuclear facility. Attention focuses on the contested nature of the evidence concerning the impact on cancer.

On March 28, 1979 there was a nuclear accident at the Three Mile Island plant near Harrisburg, Pennsylvania, resulting in a release of ionizing radiation. A presidential commission which reported shortly after the accident suggested that the maximum dose received by a person would have been less than the average annual background level, and that no detectable health effects would arise. However, public pressure led to research (Hatch et al., 1990; 1991) which sought to find out whether cancer rates near the plant had increased after the accident. In this context, proximity was defined by studying 160,000 people living within 10 miles (16 kms) of the plant. Cancer and socio-economic data were collected for 69 census tracts. Atmospheric dispersion models provided estimates of possible doses delivered to people in each of those tracts, though these estimates were very uncertain.

The initial study (Hatch et al., 1990) looked at childhood cancer, and leukaemia in particular, during both the pre-accident period (1975 to March 1979) and following the accident (up to 1985) but since there were only four cases of leukaemia, and relatively few of other cancers, little confidence could be placed in the estimates of association between cancer incidence and exposure to radiation. As far as these authors were concerned, there is no convincing evidence that radiation releases led to an increased risk of cancer. Later work by the same group detected an increase in all cancer in 1982 and 1983, but this was ascribed to factors other than radiation, such as the increased propensity, following the accident, to seek care. This research has been criticized by another group of researchers (Wing et al., 1997), who examine the association between cancer incidence and radiation dose, with controls for age, sex, socio-economic status, and pre-accident variation in incidence. Their work suggests that, following the accident, cancer incidence increased more in areas estimated to have been in the pathway of radioactive plumes than in less exposed areas. The fact that many years may need to elapse before any further increase in risk is detected, because of the long latency periods for many cancers, means that the story remains incomplete.

The Three Mile Island accident was an incident of short duration, in contrast to the longer-lasting impacts of more "chronic" air pollution or, indeed, other nuclear disasters such as that at Chernobyl in the Ukraine. One major study that illustrates the health problems of living near long-term point sources of pollution involves the Monkton Coking Works in South Tyneside, north-east England, a plant emitting smoke and SO_2. The plant has been the focus of both traditional geographical epidemiological investigation (Bhopal et al., 1994) and also analysis of popular or lay beliefs about respiratory disease, using a mix of quantitative and qualitative techniques (Moffatt et al., 1995). In both studies, three areas were chosen for detailed survey work. Two were near the coking plant, one in the immediate vicinity (the "inner" zone), the other ("outer" zone) immediately to the north (figure 7.4). The third was a control area located between six and ten kilometers from the plant. The authors show how all three zones are broadly comparable in terms of socio-economic status. Bhopal and his colleagues (1994) looked at a large variety of health outcomes, including: mortality data (both all-cause and specific causes, such as lung cancer and circulatory disease); cancer registrations (respiratory and other cancers); tests of respiratory function among children; general practitioner (GP) consultations for respiratory and non-respiratory conditions; and a community survey of ill-health. Results indicated no significant difference in adult mortality or cancers among the three zones, and rates of self-reported asthma and bronchitis were broadly similar. For self-reported wheeze, cough, and sinus trouble there was a gradient of ill-health, with higher prevalence in the inner zone, closest to the coking works. This is an interesting finding, as it suggests some impact of the pollution on the upper respiratory tract, but not the lower tract or lung. The authors also had measurements of SO_2 available to them from a small number of monitoring sites. Analyzing GP consultations by pollution level in the inner zone (table 7.5) indicated a clear and significant increase in the consultation rate for respiratory problems as SO_2 levels increase, while consultations for non-respiratory problems did not vary with pollution. As the authors note, an analysis of readily available and routine sources of data reveals much

Table 7.5 Air pollution and general practice consultations (Monkton coking works study)

Air pollution (SO₂) level	Consultation rate in inner zone	
	Respiratory conditions	Non-respiratory conditions
Level 1 (highest)	752	2182
Level 2	713	2448
Level 3	625	2361
Level 4	560	2880
Level 5 (lowest)	424	2399

Annual rates per 1,000 population
Source: Bhopal et al. (1994: 244)

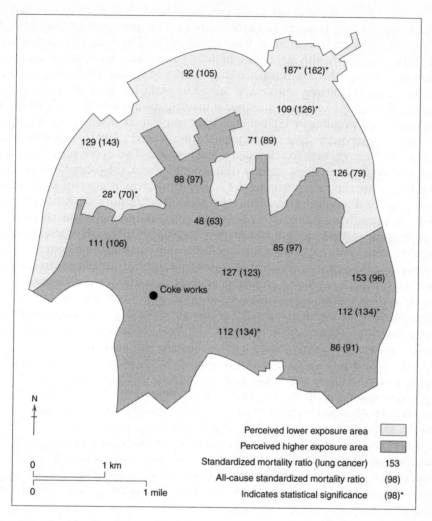

Figure 7.4 Standardized mortality ratios for lung cancer (all causes in parentheses) in relation to location of Monkton coking works (Source: Bhopal, R.S., Phillimore, P., Moffatt, S., and Foy, C. (1994) Is living near a coking works harmful to health? A study of industrial air pollution, *Journal of Epidemiology and Community Health*, 48, 237–47, reproduced with kind permission of BMJ Publishing Group)

less than one based on the harder-to-collect data on primary care consultations and self-reported morbidity. Bhopal and his colleagues conclude that "[T]he excess of respiratory problems observed in those living close to the works can best be explained as a result of their exposure to its emissions" (Bhopal et al., 1994: 246).

This study was originally sought by the local government, who, along with the coking works, had been pressed for a scientific study of the health problems perceived and suffered by local residents. These residents had their own strongly held beliefs about the link between pollution from the plant and their own experience of ill-health. Moffatt and her colleagues (1995) looked at these health perceptions via a postal survey that included both closed questions and an open section for comments. Interestingly, the three areas analyzed earlier (inner zone, outer zone, and control area) generated similar concerns about housing and family problems and difficulties with neighbours; problems that one would not expect to be associated with proximity to the coking works. On the other hand, there were clear differences among the zones in terms of complaints about dust, dirt, noise, smell, and air pollution. The quantitative evidence on health perceptions was reinforced by qualitative data. One parent of a 12-year-old commented: "If the plants and trees were dying, the air pollution couldn't have done me much good", while a 42-year-old woman reported that "I have always believed that the cokeworks were to blame for all my sinus problems as I never had any symptoms before I exchanged houses to my present address. I lived at my previous home for seven years without any sinus problems." A 53-year-old woman had "lived for many years with polluted air, the smell of sulphur carried with the wind, smoke billowing out day and night and window sills and cars covered with soot and grit" (all quotes from Moffatt et al., 1995: 888–9). The authors conclude that local anxieties are justified.

Popular ("lay") beliefs are often ignored or dismissed by public health professionals and epidemiologists as irrelevant distractions from the real task of uncovering the "facts." It is quite easy to belittle such beliefs as "sensitization bias": to suggest that those living near sources of pollution are so aware of the existence of the plant that any attempt to interview them about it will inevitably encourage them to come forward with a long list of non-specific health problems. For Moffatt et al. (1995: 884), however, "epidemiological research that attempts to put to one side local concerns and popular beliefs thereby *distorts the very reality* it so assiduously seeks to describe" (my italics). For them, public concerns are as much a source of "data" as that emerging from large-scale surveys, perhaps more so, since the latter are divorced from the social context. And the dismissal of popular beliefs as sensitization bias is readily countered by noting that the perceptions of ill-health were quite specific, and restricted to respiratory complaints. Ironically, however, when the results of the Monkton study were reported to a residents action group they, like some local government officials, felt the study to be of limited value since it "only" produced evidence of quite minor respiratory problems rather than the "serious" ill-health (such as cancer) which was of more concern.

Although of a totally different order of magnitude we can see broad parallels between studies of point sources in both the developed and developing worlds. At one level we can point to the wider structural economic forces at work, which give more weight to profit and less to community health. But we can also point to the locations of these point sources, which are, partly at least, shaped by the inability of local people to object to their being sited amongst them. Noxious facilities tend to be located in areas where political opposition is likely to be minimal. Geographers and public health professionals need to give much greater attention to issues of environmental equity and to ensure that those "point sources" most likely to impact upon health are located as far as possible from any population, including one which lacks the political "muscle" to fight location decisions.

Concluding Remarks

There are good grounds for suggesting that there are many clear links between air pollution and human health. The evidence indicating that radon gas emissions add to the burden of lung cancer, particularly among smokers, is quite convincing. Research suggests that a pollutant such as ozone triggers asthmatic attacks among those suffering chronically from the disease. Studies of the impact of vehicle exhaust emissions are more equivocal, though other literature implicates particulates and nitrogen oxides as damaging to respiratory health. Some point sources of pollution have had devastating effects, while others have been harder to assess. Partly this is because, as with all epidemiological work, studies are bedevilled by problems of confounding and exposure assessment. Aggregate studies of health outcomes in geographical areas suffer from both problems, and we need to use quite sophisticated environmental models, or better still, reasonably dense networks of monitoring sites, in order to get better characterization of exposures. This applies to all sources of pollution, whether point, linear, or areal.

Individual studies, if well-designed, can collect data on potential confounders (such as smoking) and offer a better chance of detecting impacts on health caused by the exposure (air pollution) of interest. But if studying disease with long latent periods, such as cancers, these too will be found wanting unless they can assess exposure at previous places of residence or work. Further, studies investigating mortality, or serious chronic disease, need to be set against other studies which recognize that death, or even hospitalization, is the tip of a well-submerged iceberg.

As is clear from most of the studies reviewed in this chapter, research has tended to adopt a classic positivist stance. Apart from work by Moffatt and her colleagues (1995), few studies look at lay beliefs about sources of air pollution and their effects. Classically trained epidemiologists will dismiss the value of this, pointing to sensitization bias and arguing that those living near possible point sources of pollution will, literally, see the source and be keen to attribute ill-health to such sources. Yet, as Moffatt and her colleagues suggest, it is surely misguided to ignore those with the most immediate experience of pollution, those with direct exposure.

I think therefore, that while there must continue to be well-designed epidemiological studies, ideally of a longitudinal nature (examining health outcomes before and after pollution incidents or episodes) we must also pursue other approaches to scientific investigation. Some of these need to be informed by a better understanding of the views of those affected, or potentially affected, by pollution sources. Equally, there is much to be gained from pursuing more structuralist accounts, ones that recognize the deeper structures underlying associations between air pollution and health, including the interests of business that may put profit before occupational and community health. We need to add a social and political model of health to classical epidemiological accounts.

FURTHER READING

For general background reading on air pollution and health see Ayres (1997). Dunn and Kingham (1996) discuss some of the methodological difficulties in assessing the links between air quality and health.

The journal *Health Physics* is a good source of material on the links between radon gas and human health. For an overview of work on traffic and health see Whitelegg (1997, Chapter 10) and McCarthy (1999).

Full accounts of the Bhopal tragedy are to be found in Shrivastava (1992) and Bogard (1989). The former gives more information on the health consequences, while the latter is a stunning interpretation of the politics and political economy of risk and hazard as exemplified by the Bhopal disaster. An account of the Chernobyl disaster and its consequences is in P.R. Gould (1990) *Fire in the Rain*, Polity Press.

For a very clear argument in favour of a social interactionist perspective on air pollution see Phillimore and Moffatt (1994).

Proximity to hazardous waste sites can expose local populations to both air and water pollution. For a comprehensive picture of the public health impacts of such sites, primarily from a US perspective, see National Research Council (1991). This also has valuable general material on the principles of environmental epidemiology.

CHAPTER 8

WATER QUALITY AND HEALTH

We looked briefly in Chapter 5 at the important contribution of safe, clean water, and sanitation improvements, to health outcomes. I want now to look in more detail at links between water quality and health. Such links are particularly acute in large parts of the developing world, and so I give some attention first to two of the most serious waterborne diseases occurring in these areas, cholera and schistosomiasis. I turn next to other gastrointestinal diseases, some of which are of increasing importance in the developed world, before looking at the contamination of water by chemicals and the possible health consequences of this. The hardness of water has been the focus of some research attention, and I devote some space to this topic. The contamination of water supplies by hazardous wastes has generated considerable public concern and research effort and this too is considered in the present chapter. Last, although the burden of disease and illness pales into insignificance compared with some of the health risks considered earlier, I examine some of the research evidence pointing to associations between morbidity and exposure to bathing waters.

Most of the work I examine is drawn from the environmental epidemiology research literature. Much of this is necessarily "positivist" in orientation, seeking to test hypotheses concerning exposure and outcome. I hope it is clear that there is a role for other perspectives, and that, as in the previous chapter, it becomes clear that we should not neglect the individual's perception of risk. Most important, I am increasingly persuaded that we cannot secure a full understanding of the associations between disease/illness and water quality, without addressing wider structural and societal determinants.

Waterborne Diseases

Cholera

Cholera is a disease caused by the ingestion of a bacterium, *Vibrio cholerae*, present in water contaminated by faecal matter; the course of the disease is described graphically in Watts (1997: 173). There are over 100 serotypes of *V cholerae*, two (O1 and O139) being responsible for the major disease epidemics. Symptoms are a massive loss of water and salts, leading to watery diarrhoea. If the depleted body fluids are not replenished (which is possible via the kinds of oral rehydration therapy reviewed in Chapter 5) then kidney and heart failure are a consequence.

I begin by examining some of the historical, as well as contemporary, research on cholera, which, as Hunter (1997: Chapter 12) notes, was the first disease shown by epidemiological methods to be waterborne. There are two reasons for considering cholera here. First, the person who established the link between cholera and contaminated water, John Snow, produced a study in 1854 that has assumed almost mythological status as a piece of geographical epidemiology. This work by Snow, followed later by the microbiological research of Robert Koch that implicated the bacterium, played a major role in producing the sanitary reforms which helped control the disease in the developed world. The second reason for an extended discussion of cholera is quite simply because it continues to be a major cause of mortality across the globe.

The bacterium *V cholerae* survives best in moderately saline waters, and thus prefers estuarine conditions, ideally where the temperature remains at 10°C (50°F) for several weeks at a time. Since it resides in the gut of apparently healthy individuals who may be traveling widely, it can spread via other transport routes. Thus French soldiers working in early colonial Algeria transported the disease there in the early 1830s (Watts, 1997: 172). Britain suffered five cholera epidemics during the course of the nineteenth century, killing an estimated 130,000 people (see Learmonth, 1988: 143–52 for maps of its diffusion in the early- to mid-nineteenth century). Over the same period India lost about 25 million people from the disease.

Before 1817 large parts of India were thinly populated by semi-settled and nomadic pastoralists in an "ecologically sound, cholera-free rural order" (Watts, 1997: 181), but British policies of compulsory resettlement onto fixed village sites disrupted this balance and did little to mitigate the effects of poor harvests. Malnutrition contributes to a pre-disposition to cholera since the body's defence mechanisms are weakened. The major epidemic of 1817, which originated in Jessore in Bengal (now Bangladesh) and spread to central India (helped on its way by the movements of the East India Company army), was preceded by famine. In the later-nineteenth century, other human intervention aided the spread of the disease, not least the role of the Public Works Department in overseeing famine relief by concentrating huge numbers of people into camps supplied with contaminated drinking water, and by constructing a network of railways and irrigation canals unaccompanied by the drainage ditches needed to remove surplus water. "Impacting most heavily on the ability of cholera to slaughter millions were gentlemanly capitalists' very substantial investments in irrigation, in railways and port facilities and, equally decisively, the near absence of investment in public health" (Watts, 1997: 168). Watts also points a finger at the role of medical professionals who did little to halt the spread of cholera among the native Indian population, while for others the disease "was associated with much that European medical officers and administrators found

Table 8.1   Cholera in 1999

Country	Number of cases
Mozambique	44329
Malawi	26508
Nigeria	26358
Afghanistan	24639
Somalia	17757
Democratic Republic of Congo	12711
Tanzania	11855
Zambia	11535
Kenya	11039
Madagascar	9745
Ghana	9432
Zimbabwe	5637
Uganda	5169

Source: World Health Organization (2000)

outlandish and repugnant in Hindu pilgrimage and ritual – so much so that the attack on cholera concealed a barely disguised assault on Hinduism itself" (Arnold, 1988: 8). And while some degree of medical intervention was essential in order to extract maximum output from the workforce – providing this did not eat into profits – this served mainly to bolster capitalism's enterprise.

Following this first recorded pandemic in the early-nineteenth century (1817–23), others followed frequently into the early part of the twentieth century. All originated in the Ganga (Ganges)-Brahmaputra delta in what is now Bangladesh and spread throughout much of Asia, the Middle East, Africa, and the Americas. In the early 1960s a new strain (El Tor) emerged from Indonesia before spreading world wide. More recently, the O139 strain originated in India in 1991 and has spread into south-east Asia. In 1994 nearly 11,000 deaths and nearly 400,000 cases were reported to WHO, though these figures must be treated with caution as there is likely to be massive under-reporting; publicizing an outbreak of cholera is unlikely to do much for fragile tourist industries. Those countries with more than 5,000 reported cases of cholera in 1999 are listed in table 8.1. Of these, all but one is in Africa.

There have been a number of studies in high-incidence areas, such as Bangladesh. These show that the disease is socially patterned, with better-off families having lower disease incidence. In addition, higher incidence has been reported close to a hospital dealing with the disease, since local residents have used canal water polluted by effluent discharged by the hospital into the canal (Levine et al., 1976). Studies in South Africa have confirmed that incidence is elevated in areas of lower social status, while work in Mali found that those drawing water from one well located within 10 meters (three yards) of three pit latrines were more likely to contract the disease (Hunter, 1997: 110). Cholera appeared in Latin America during the 1990s, killing over 11,000 people in the first five years of the decade. It first appeared in a coastal village in January 1991 but only five weeks later had diffused to 24 out of 29 Peruvian departments (Mintz et al., 1998: 86). In Mexico, the spatial patterning shows a north-south gradient, with higher incidence in the poorer southern states, while in Brazil it is the poorer northern states where incidence is highest (Mintz et al., 1998: 76). While it may well be the case that "health education and simple personal hygiene such as washing with soap can be significantly protective" (Hunter, 1997: 114) it is surely the case that to halt disease incidence and spread requires an attack on underlying structural determinants of poverty and poor sanitary infrastructure. As Mintz and colleagues argue, "the crowded populations of the developing world's periurban slums . . . still bear the burden of cholera disease" (Mintz et al., 1998: 65). WHO denoted the 1980s as the Drinking Water and Safe Sanitation Decade, but the provision of basic needs had been countered by rapid population change, to the extent that 2.25 billion people do not have access to safe drinking water, and 3 billion lack access to basic sanitation (Mintz et al., 1998: 80).

From an historical geographical perspective there have been a number of interesting studies of cholera. For example, Howe (1972) mapped the spread of the second pandemic (1826–51) from India to Britain, and within Britain (see also Learmonth, 1988: 143–8). The likely route was from India to Russia and then, via the Baltic port of Danzig to Sunderland in October 1831. From there it spread both north into Scotland and south to Yorkshire, the Midlands, and then south-west England. A spatial diffusion study by Pyle (1969) explored three epidemics (1832, 1849, and 1866) in the nineteenth-century United States (see also Cliff et al., 1998: 272–3). Pyle was particularly interested in the relative importance of contagious and hierarchical diffusion processes. He showed that, in 1832, spread was dictated by distance (contagious diffusion), with towns near the origin of the outbreak registering the disease much earlier than those far away. In 1849 and 1866 improvements in transport meant that towns far from the origin in terms of geographical distance had become better connected in terms of travel time; spread was more hierarchical, with larger towns and cities reporting the disease much earlier than smaller settlements near the origin.

The classic epidemiological work – which shows a geographical imagination – is, as noted earlier, due to Dr John Snow working in London in the middle of the nineteenth century as the second cholera pandemic was under way (Cliff and Haggett, 1988: 7–11). In the second edition of his book, *On the Mode of Communication of Cholera*, published in 1854, Snow produced a map which showed that most of the deaths from cholera were to be found in a small area of Soho, and were of people who had taken water from a pump located in Broad Street; the water had become contaminated by a leaking cesspool. Snow saw the spatial patterning and its relationship to proximity to the pump: "the deaths are most numerous near to the pump where the water could be more readily obtained. The wide open street in which the pump is situated suffered most, and next the streets branching from it, and especially those parts of them which are nearest to Broad Street" (quoted in Cliff and Haggett, 1988: 53–5).

Accounts of Snow's work appear in numerous books on epidemiology and medical geography and focus on the "detective" work that led to the postulating of waterborne transmission (see Paneth et al., 1998 for further background). But as Snow himself made clear, this is only part of the story and links to material circumstances and the structural shortcomings of some of the contemporary water companies are also relevant. Snow (cited in Cliff and Haggett, 1988: 7–8) indicated that cholera in London "has borne a strict relation to the nature of the water supply of its different districts, being modified only by poverty, and the crowding and want of cleanliness which always attend it." Cholera was associated in particular with two companies drawing water directly from the River Thames, a "kind of prolonged lake . . . receiving the excrement of two millions and more inhabitants" (Snow, in Cliff and Haggett, 1988: 8). A contemporary cartoon (reproduced in Cliff and Haggett, 1988: 9) indicated where blame lay; against a drawing of industrial effluent was the following verse:

> These are the vested int'rests, that fill to the brink,
> The network of sewers from cesspoool and sink,
> That feed the fish that float in the ink-
> -y stream of the Thames, with its cento of stink,
> That supplies the water that John drinks.

> (reproduced with kind permission of Professor
> Andrew Cliff from Cliff and Haggett, 1988: 97)

Schistosomiasis

Schistosomiasis (also known as bilharzia) is a family of diseases resulting from infection with flatworms or "flukes" of the genus *Schistosoma*. The life cycle involves three factors: the human host, the worm, and an intermediate host, water snails (genus *Bulinus*). The worm's eggs are released via the urine or faeces of an infected person into freshwater, where the eggs hatch and penetrate the snail. After four to six weeks the cercariae (the worm's larvae) are released into the water and penetrate the skin of a human host. The worms migrate to the liver or lungs and, once mature, to the veins of other body organs. Different species of *Schistosoma* affect different parts of the body, so that some cause bladder and kidney disease, others disease in the liver and intestines. As is clear from this brief description, the water snail is a crucial element. This requires warm, preferably still or slow-moving and muddy, freshwater, with optimum temperature in the range 20–30°C (68–85°F).

Schistosomiasis presently affects over 200 million people world wide and although mortality is relatively modest (20,000 deaths per annum, due to subsequent development of cirrhosis and bladder cancer) it is a very debilitating disease in the tropics, second only to malaria in terms of prevalence (Learmonth, 1988). Swimming, playing in the water, and fishing all provide opportunities for disease transmission, and this explains the higher incidence among children and young adults. More than 80 percent of those infected are in sub-

Saharan Africa. Treatment is via chemotherapy in the human host, though mol-
luscicides can control the spread of the snail. However, as we saw in earlier
chapters, diagnostic procedures and safe drugs do not necessarily reach those
most in need.

Explanations of disease incidence reveal the classic tension between behavioral and structural determinants. Given that the disease cycle has a clear link
to poor disposal of faeces and urine, one argument for disease control would
be for better education about hygiene. On the other hand, it is clear that
alterations to water-regimes have added to health hazards in some areas. While
major irrigation schemes and flood control measures have had positive outcomes
(in terms of increased food availability and therefore improved nutrition) they
offer new habitats for the snail hosts and therefore new opportunities for the
disease to spread. Such effects were recognized as long ago as 1971, when WHO
suggested that schemes such as Aswan High Dam on the river Nile (and the
large Lake Nasser behind it) were compounding the problem of infection.
For example, one species of the worm, *Schistosoma mansoni*, affected only 3
percent of the population of one nearby Egyptian village in 1935, but this had
risen to 73 percent by 1979, an increase that Abdel-Wahab et al. (1979) attributed to the dam construction. More recently, the construction of the Diama
dam on the Senegal River has introduced intestinal schistosomiasis to Senegal
and Mauritania.

Geographical research on schistosomiasis illustrates nicely the range of
approaches to the geography of health reviewed in Chapter 2, including broadly
positivist, spatial analytic work and more sociologically informed, qualitative
research. For example, there is a wealth of research on the spatial distribution
of infection at detailed spatial levels, research that examines the relationship
between intensity of infection (egg counts) and a range of explanatory variables,
including distance from water sources; predictably, distance to water sites is
inversely associated with infection, though the strength of this association varies
among study areas (Kloos et al., 1997). Such research uses both detailed field
work, as well as GIS and remote sensing (see Malone et al., 1997, for example),
to map infection and to suggest areas in which control strategies should be
focused. Others, however, prefer much more intensive and qualitatively
grounded approaches. For example, earlier research by Kloos et al. (1983), some
of whose detailed maps are reproduced in Learmonth (1988: 245–9), used an
ethnographic approach to observe and record the behavior of boys aged between
five and 16 living in a small village, El Ayaisha, in upper Egypt. Home addresses
were mapped, as were trips made to bathing places (usually the River Nile, but
occasionally to canals), but time and duration of bathing, together with proportion of body immersed, were also noted, as were the kinds of activity undertaken (washing, swimming, fishing, and so on). Detailed maps of egg counts
detected in urine samples were also prepared. Swimming seemed to be the pre-
dominant source of exposure to infection. Watts et al. (1998) adopt the framework of time geography, and methods of participant observation, to explore
the detailed daily use of water sites by one household in Morocco. They argue
too for a more sensitive awareness of the gendered use of space and water resources, suggesting that "this strategy will develop insights which would not be

yielded by a conventional water study which looks at individual water contact activities dislocated from their social and cultural context" (Watts et al., 1998: 756).

The deeper underlying causes of the disease are poverty and poor living conditions. Hunter (1997) reports on a Brazilian study which found the highest egg counts in faeces from children of servants rather than children of household heads, and that low family income was also associated with high egg counts. Those households with flush latrines had lower egg counts than those with only pit latrines or none at all. Population displacement due to the movement of refugees has also played a part; for example, political instability in the Horn of Africa has led to the disease spreading to Somalia. This factor, coupled with the need for resources to be directed at the underlying causes of the disease (poverty, and lack of infrastructure) suggests that explanations of disease prevalence, and solutions to the problem, demand a structural or political economy approach.

Gastroenteritis

Three common sources of gastroenteritis around the world (though much studied particularly in the developed world) are due to the protozoan parasites *Giardia* and *Cryptosporidium* and the bacterium *Campylobacter*. In all cases, the main symptom is diarrhoea, while animal and human sewage are key sources of environmental contamination. Giardiasis is quite widespread throughout the world and is a common cause of diarrhoea among international travelers, particularly among those who have consumed local tap water. The parasite survives in unfiltered and unchlorinated water supplies and therefore some outbreaks are due to the consumption of untreated water, such as that from streams (which may be contaminated by animal waste). Hunter (1997, Chapter 7) reports on a number of cases in the literature. In other instances, local water supplies are accidentally contaminated by overflows of sewage. Cryptosporidiosis outbreaks have, since the 1980s, been responsible for a considerable disease burden in the developed world. As with giardiasis there is a close correlation between attacks of diarrhoea and the consumption of contaminated tap water, such contamination again arising from flaws in filtration systems in water treatment works. One of the largest outbreaks was in Milwaukee in 1993, when over 400,000 people were affected. Here, water is drawn from Lake Michigan, and although this is treated by chlorination, filtration and other procedures it proved difficult to identify the precise area of system breakdown (Mackenzie et al., 1994). In other cases, consumption of groundwater, possibly contaminated by infected cattle, is a likely route for disease transmission.

Campylobacter transmission occurs though the consumption of undercooked meat (especially chicken) and untreated (unpasteurized) milk, as well as via contaminated water. Several studies implicate unboiled tap-water or groundwater supplies as sources of infection. Reservoirs that are open to animal wastes, or storage tanks that may be contaminated by roosting birds, are possible sources. The problem is particularly acute in rural areas. For example, those living in a rural community in Norway developed severe gastroenteritis following consumption of tap water fed by upland lakes: during spring, snow-melt flooding washed sheep faeces into the lakes from a path running alongside (Hunter, 1997: 140). The research evidence on campylobacter comes predominantly from the developed world. This might lead to the conclusion that the problem is unknown, or limited, in the developing world. This is grossly mistaken, as some studies (Hunter, 1997) indicate. For example, campylobacter is a major source of diarrhoeal disease in sub-Saharan Africa, again due mainly to the absence of a proper water supply.

There is little doubt that explanations for geographically localized outbreaks of waterborne disease must be sought not at the individual, but at the macro, level, a conclusion endorsed by Bradley (1998) who argues that diseases such as giardiasis and cryptosporidium have emerged because of inadequate water treatment. He sees the pressures to reduce costs and to increase competition as symptoms of a culture that is prepared to tolerate violations of public health in order to secure cheaper water (or, in other cases, food: see box 3.1, page 57).

Chemical Contamination of Drinking Water

Aluminium, Fluoride, and Arsenic

One of the most widely publicized public health incidents in Britain over the past 20 years was the accidental contamination of the local water supply in Camelford, Cornwall (south-west England). Here, in July 1988, a lorry driver tipped 20 tons of aluminium sulphate into a tank feeding the Lowermoor waterworks, exposing an estimated 20,000 residents to high levels of aluminium (and also lead and sulphate). Two reports were produced by a panel of scientists, chaired by Dame Barbara Clayton (Williams and Popay, 1994; Hunter, 1997: 246–8), and these concluded that while there were short-term effects on health there were no long-lasting consequences.

Residents continued to complain, however, 11 years after the incident. As one man in his 70s recalls (the *Guardian* newspaper, June 9, 1999): "We lose things or put them down in funny places, something we never did before. It's got to the state that I can't even repair the car. I pick up the manual and read it but by the time I have got the bonnet up I've forgotten what I have read." A critic might well counter that this is not particularly unusual in a 70-year-old; however, the same man reported that shortly after the original incident: "I was sitting in a chair and found I couldn't move. My wife's hair turned red when she washed it and she had a rash up her arms. We had nausea, diarrhoea and mouth ulcers."

The Clayton Committee suggested that "it is not possible to attribute the very real current health complaints to the toxic effects of the incident, except inasmuch as they are the consequence of the sustained anxiety naturally felt by many people" (cited in Williams and Popay, 1994: 106). This is reminiscent of the dismissive "sensitization bias" in the Monkton coking works study of air pollution (Chapter 7). However, a recent study by Altmann et al. (1999) sug-

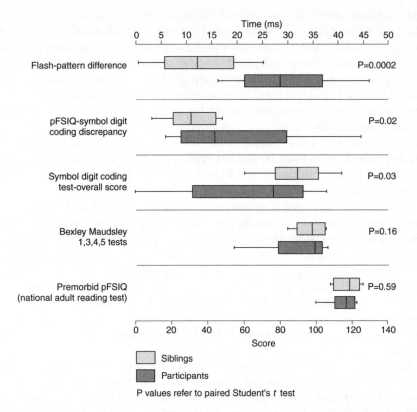

Figure 8.1 Comparison of cases (participants) and controls (siblings) in psychological tests following the Camelford water pollution incident (Source: Altmann, P., Cunningham, J., Dhanesha, U., Ballard, M., Thompson, J., and Marsh, F. (1999) Disturbance of cerebral function in people exposed to drinking water contaminated with aluminium sulphate: retrospective study of the Camelford water incident, *British Medical Journal*, 319, 807–11, reproduced with kind permission of BMJ Publishing Group)

gests that this "scientific" conclusion may be premature. The authors studied 55 people who were considering legal action and who had claimed to suffer organic brain damage. They also studied siblings of the cases, people who were broadly similar in age and who lived away from the Camelford area. A set of psychological tests to establish attention, coordination, visual skills, and memory were used to compare the cases and controls. A sensitive test for organic brain disease, the "symbol digit coding test," reveals that the performance of the cases is significantly worse ($p = 0.03$) than that of their siblings; box-plots revealing the lower and upper quartiles demonstrate this well (figure 8.1). The ability of cases (participants) to respond to visual stimuli is even worse ($p = 0.0002$). The authors conclude that "there are no other known causes for the effects that we have described in the people from Camelford" (Altmann et al., 1999: 810).

This example serves as a classical illustration of the tension between a "scientific" and a "lay" account of a pollution incident. Here, unlike in many cases, there is no disputing that pollution occurred. The debate centers round its health impacts. A scientific study published more than 10 years later points to significant psychological morbidity. Until this was published, the scientific evidence was weak and "experts" pointed to dangers of reporting and sensitization bias; if people know they are likely to have been exposed they are more likely to report symptoms. Should we have accepted the scientific account of the expert (Clayton) committee, or accounts of people who clearly were exposed to high levels of aluminium? Altmann and his colleagues have provided new evidence for accepting a link between exposure and morbidity, though this evidence remains contested.

There have in fact been concerns for many years of possible associations between raised levels of aluminium and the risk of neurological disease, including dementia, Alzheimer's disease, and motor neurone disease (amyotrophic lateral sclerosis). One geographical study in England (Martyn et al., 1989) considered rates of Alzheimer's disease among the population aged 40–69 years, in 88 local authority districts. Water quality data were obtained from the records of water companies. The risk of disease was 1.5 times higher in counties where aluminium levels were 0.02–0.04 mg/l, although at greater concentrations (>0.11 mg/l) the risk was no higher. The difficulty with this study, as with all aggregate studies, is that exposure is poorly characterized; can we really believe that all individuals currently living in a large area have had identical exposures to water of a particular chemical composition? A case-control study (Forster et al., 1995) in northern England failed to find a significantly elevated risk of Alzheimer's disease among those exposed to levels in excess of 0.15 mg/l. It remains difficult to know whether aluminium exposure "causes" these diseases or whether aluminium concentrations tend to increase more generally with aging.

The example of fluoride illustrates neatly some issues concerned with risk perception and risk communication. We saw in Chapter 4 how dental health among children improved in areas having fluoridated water, although material deprivation was also a factor. Fluoride reduces the solubility of tooth enamel and thus offers protection against decay. For example, a South African study (du Plessis et al., 1995) indicated that there was a 50–80 percent reduction in tooth decay among black children when fluoride levels were between 0.2 and 0.9 mg/l. Evidence of this sort has led to calls from community dentists and public health professionals to have fluoride added to water supplies lacking the chemical. In Britain, the USA, and elsewhere there are vociferous groups who oppose such fluoridation; some cite evidence of links between fluoridation and cancer (bone cancer, or osteosarcoma, in particular), while others oppose a public health intervention over which they have no choice. One has some choice, they argue, over whether to exercise, and how much fatty food to consume, but fluoridated water is hard to avoid. Some cases of accidentally raised levels of fluoride (for example, due to faulty equipment) have produced acute episodes of gastrointestinal illness and nausea. Moreover, some studies in China, Kenya, Japan, and elsewhere indicate that where concentrations of natural fluoride are high (above 2 mg/l) there is a risk of fluorosis; here, calcium in bones is replaced by fluoride and bones begin to soften and crumble (Hunter, 1997: 257–8).

The association between fluoridated water and cancer was given publicity by a national newspaper (the *Guardian*, June 1999), which claimed that "in fluoridated areas of America, bone sarcoma rates among boys aged 9 to 19 are between three and seven times higher than in non-fluoridated areas." The newspaper article failed to point out that this study looked only at seven counties in one state, New Jersey, and simply compared aggregate rates of bone cancer in fluoridated and non-fluoridated communities. Indeed, osteosarcoma was identified in 12 males aged under 20 years in fluoridated communities, and only eight in non-fluoridated areas. This is hardly compelling evidence in favor of an association between exposure to fluoride and bone cancer. Other work has found that while osteosarcoma rates are higher in fluoridated areas, this bears no relation to time of fluoridation. Clearly, this level of detail cannot be discussed in a newspaper article, but for those reading it a doubtful association between "cancer" and fluoridated water is now embedded in their consciousness. For a full review of epidemiological studies see Cook-Mozaffari (1996).

Arsenic is a highly toxic metal and its presence in groundwater means that those consuming such water are at risk of serious chronic disease. There are high natural concentrations of arsenic in water drawn from aquifers in the Ganges delta and this has led to serious skin conditions ("arsenical dermatitis"; see Chakraborti and Saha, 1987). In south-west Taiwan the incidence of "blackfoot disease" (a peripheral vascular disease affecting the feet but sometimes hands, and leading to gangrene) is high and has been linked firmly to the consumption of groundwater contaminated by arsenic. Chen et al. (1988) reported results of a case-control study of over 300 patients which showed that those exposed to such water for over 30 years were 3.47 times more likely to develop the disease than those unexposed. Other research in Taiwan indicates that there are high correlations between arsenic levels in water and the incidence of lung, skin, kidney, and bladder cancer (Wu et al., 1989). Concentrations of above 0.3 mg/l seem to be a risk factor for these cancers and other chronic disease. Thus Hunter (1997: 255) concludes that "arsenic in drinking water other than at very low concentrations has a significant adverse health impact." One solution to the problem is emerging from geochemists, who suggest that a relatively simple device can be used to bind iron hydroxide and arsenic; this can then be filtered out. The system is now being adopted in Bangladesh.

Water Hardness

Water hardness (of which calcium and magnesium are the main components) is thought to be implicated in the development of eczema, and a study of children aged between four and 16 years in Nottingham, England, used a GIS approach to offer some confirmation (McNally et al., 1998). Childrens' residential addresses were assigned to one of 33 water supply zones, for each of which data

Table 8.2 Water hardness and prevalence of eczema among children under 11 years in Nottingham, England

Water hardness[a]	Lifetime reported eczema prevalence (%)	Odds ratio[b] (95% CI)	1 year reported eczema prevalence (%)	Odds ratio[b] (95% CI)
118–35	21.2	1.00	12.0	1.00
151–7	23.4	1.13 (0.88–1.45)	14.1	1.19 (0.88–1.62)
172–214	22.7	1.16 (0.92–1.46)	14.6	1.32 (1.00–1.76)
231–314	25.4	1.28 (1.04–1.58)	17.3	1.54 (1.19–1.99)

[a] In mg/l of calcium and magnesium salts
[b] Adjusted for age, sex, socio-economic status, and distance from nearest health center
Source: McNally et al. (1998: 529)

on water hardness (divided into four categories) were obtained. After adjustment for potential confounders there was evidence of an elevated risk of eczema among younger children (under 11 years), whether reported within the past year or at any time in the child's life (table 8.2). The odds of reporting eczema within the previous year was 1.54 in the areas with the hardest water (that is, 54 percent higher than in areas with soft water). A possible explanation is that calcium in the water acts as a skin irritant; in addition, hard water makes it more likely that there is a greater need for soap and shampoo when bathing, in order to develop a sufficient lather.

In other contexts, harder water seems to be protective of health. A well-known example is cardiovascular disease, where the British Regional Heart Study (Pocock et al., 1980; 1982) suggested that cardiovascular mortality in soft-water areas was about 15 percent higher than in harder-water areas. Research in the USA, South Africa, and other parts of Europe supports this finding. As figure 8.2 demonstrates, the unadjusted SMR in areas of Britain served by supplies of soft water is about 120, but only 90 in harder water areas, and the gradient, while attenuated, remains after adjustment for climate and socio-economic status. Such adjustment is critical, since areas of high mortality tend to be in the north of the country, where material deprivation (very broadly speaking) tends to be worse. There are, however, difficulties in interpreting this finding, because water hardness is correlated with other water parameters, including nitrates. As the researchers observe, "whether the lower CVD rates in hard water areas might be due to the beneficial effect from bulk minerals (e.g. calcium and magnesium) or the absence of harmful minerals or trace elements (e.g. sodium, lead and cadmium) remains unknown" (Pocock et al., 1982: 321). Hunter (1997) suggests that soft water is more likely to corrode, and to dissolve metals such as lead and copper (used in water pipes) that may themselves be implicated in heart disease mortality.

Other difficulties, both conceptual and statistical, in assessing this relationship are considered in Jones and Moon (1987: 134–40). One is that the work reported above is an aggregate study; do the findings hold at the individual level?

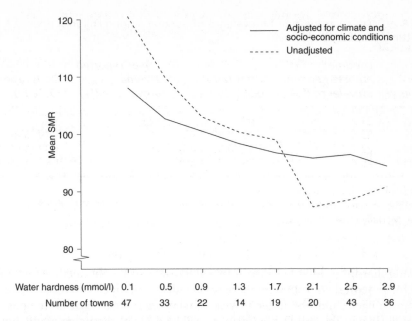

Figure 8.2 Relation between standardized mortality ratio for cardiovascular disease and water hardness, in a sample of British towns (Source: Pocock, S.J., Cook, D.G., and Shaper, A.G. (1982) Analysing geographic variation in cardiovascular mortality: methods and results, *Journal of the Royal Statistical Society, Series A*, 145, 313–41, reproduced with kind permission of Blackwell Publishers Ltd)

Work by Downing and Sloggett (1994) which studies premature mortality (all causes, for those under 65 years) demonstrates a significant relationship for men, but not for women. Adjusting for individual socio-economic circumstances, men living in hard water areas have a 10 percent reduction in mortality risk compared with those in soft water areas. However, the influences of unemployment, rented accommodation, and lack of a car increase the risk of premature mortality much more than does water hardness, a finding that will not surprise readers of Chapter 4.

Other Forms of Contamination

Hazardous Waste Sites

Consider now the health risks of exposure to contaminated drinking water, where the source of such contamination is sites disposing of hazardous wastes. This is not to deny that other sources of potential problems are air pollution from, or perhaps via the consumption of food supplied from areas polluted by, such sites. Of particular interest are organic contaminants, including pesticides, solvents, and petroleum products, but organics such as trichloroethylene (TCE) and trihalomethanes (THMs) have been studied in detail.

TCE is used in dry-cleaning and as an industrial solvent, while THMs are a by-product of chlorination, formed by the action of chlorine on organic compounds.

While associations between hazardous waste disposal and human health are of global concern, great attention has been paid to this issue in the USA, where the Agency for Toxic Substances and Disease Registry (ATSDR) has a remit to consider the health effects of living near hazardous waste sites. In the USA as a whole, 50 percent of the population consumes groundwater, a figure that rises to 95 percent in rural areas. Given that the US Environmental Protection Agency (EPA) has estimated that 40 million people live within four miles (6.4 km) of "Superfund" sites (box 8.1) and four million within a mile of such sites (National Research Council, 1991) there is a clear geography to the possible association between exposure and health outcome. We need to bear in mind, as we consider some of the literature, that geographical proximity does not necessarily equate to direct exposure.

Many individual sites have been the focus of investigation. For example, Lagakos and colleagues (1986) carried out research on the town of Woburn in Massachusetts, where TCE and other organics had contaminated two wells. Twenty cases of childhood leukaemia were recognized in the study area, compared with the 9.1 which would have been expected on the basis of national rates. Love Canal, in upper New York State, is one of the best known, and most intensively studied, waste sites in the USA. Here, water contaminated by benzene and other chemicals seeped into basements (and therefore exposure was by inhalation rather than direct consumption). One health outcome of interest here has been low birth weight (Vianna and Polan, 1984), results suggesting that between 1940 and 1953, when large volumes of chemicals were dumped in the site, there was a significantly increased number of infants of low birth weight in the most exposed site. Here, "exposure" was represented in terms of whether the house was on a low-lying natural drainage depression, or swale, in which

---

BOX 8.1   *Superfund*

Growing concern about hazardous waste sites in the USA led to legislation to clean up the worst of those sites. The Superfund programme was established in 1980 to locate, investigate, and clean up such sites. It is administered by the US Environmental Protection Agency (EPA). EPA conducts tests of water, soil, and air samples in order to assess the risk posed by a site. Sites are scored according to the risk they pose to public health. The worst such sites are put onto a National Priorities List and are eligible for a long-term program of remediation. This remedial action is funded using sums demanded from the companies responsible for the site contamination.

Lists of sites, together with maps, are available via the EPA's web-site (see the Appendix p. 261).

water was more likely to percolate into the ground. Other work by Paigen and colleagues (1987) has monitored the growth of children born, and living at least 75 percent of their lives, in the Love Canal area, demonstrating that such children were significantly shorter than control children, differences that remained even after adjustment for possible confounding variables.

Others have looked at associations between exposure to contaminated water and more serious adverse pregnancy outcomes, such as congenital malformations. In particular, a number of studies (see National Research Council, 1991: 191–5 for a review) have examined links between exposure and heart malformations. For instance, Goldberg et al. (1990) studied heart malformations in the Tucson Valley, Arizona, part of which had been contaminated with trichloroethylene between 1950 and 1980. The study found that 35 percent of mothers with malformed infants had conceived their child, and lived during the first trimester (three months) of their pregnancy, in the contaminated area, three times as many as those without contact with polluted water. After the contaminated wells closed, in 1981, the odds ratio dropped from three to nearly one. Animal experiments confirm the plausibility of this relationship. Assuming one can be confident that diagnoses, and record-keeping, have not changed over time, the kind of longitudinal work illustrated by Goldberg and colleagues has clear advantages over cross-sectional studies, since one can relate exposures during particular time periods to outcomes during the same, or later, periods.

Griffith et al. (1989) conducted a large-scale study of nearly 600 hazardous waste sites, spread across 339 counties throughout the USA. County-level mortality data were examined for 13 types of cancer and the rates in the 339 counties compared with 2,726 "control" counties that did not have evidence of contaminated water. The authors reported significant excesses of deaths in the "case" counties, for both men and women, from cancers of the lung, bladder, stomach, colon, and rectum. However, this was a study of aggregate areal units that did not adjust for individual-level risk factors or confounders; for example, what evidence is there that smoking is more prevalent in the "contaminated" counties, which might have accounted for the excess risk of several cancers?

There are therefore several difficulties in trying to assess the association between hazardous waste and health outcomes. Most serious is the question of exposure: how do we measure this, and to what extent can we assume that the exposure has not varied over time? What is known about the residential histories of those affected, or other possible sources of exposure, as in the workplace? Definitions of study areas and study populations are also problematic; certainly, the first of these is inherently arbitrary. Studies of areal units may be suggestive, but detailed case-control studies are needed for a more convincing result, not least because such studies can adjust for the influence of confounding variables. Further, we are typically dealing with quite rare diseases of low incidence, and epidemiological studies require larger numbers of cases in order to achieve statistical "power." All these issues, and more, are the very stuff of environmental epidemiology and have to be addressed by positivist approaches to the problem. Quite separate from these methodological issues, however, are more

> ## BOX 8.2   *Lay epidemiology*
>
> As we have seen in both this chapter and the previous one, there is a tension between "scientific" investigations of disease "clusters" and the understandable concerns of ordinary people who either live near suspected sources of pollution or claim to have detected unusual local aggregations of disease or illness. In a sense, this is a conflict between "positivist" and "social interactionist" views of the world (Chapter 2). There are numerous examples of communities claiming cancer "clusters" which public health specialists may or may not choose to investigate.
>
> Some health professionals have shown how lay people can themselves undertake "scientifically rigorous" studies in order to persuade others that there are justifiable local health concerns. Marvin Legator and his colleagues collaborated on a guide to assist local communities in investigating suspected environmental hazards, such as hazardous waste sites. The book considers issues of experimental and questionnaire design, data analysis, and how to seek assistance from other agencies. It is an excellent resource for anyone, student or concerned citizen, who wishes to engage in their own epidemiological investigation.
>
> See Brown (1992) and M.S. Legator, B.L. Harper, and M.L. Scott (1985, eds.) *The Health Detective's Handbook*, Johns Hopkins University Press, Baltimore, USA.

fundamental questions of epistemology. By this I mean once more the primacy of this kind of "scientific" knowledge over the accounts of illness given by those living near such sites. It is easy to make accusations of "scaremongering" and bias among local populations, yet it is, quite understandably, difficult to persuade those living near to contaminated sites that three cases of a rare childhood cancer might be a statistical chance fluctuation rather than having a causal link to waste disposal. There is a deep irony in the fact that we need plenty of cases of disease to secure a respectable sample size, while also wishing to minimize the burden of any life-threatening condition on worried local communities (box 8.2).

Bathing Waters

In this section I consider some of the evidence on health risks from bathing or swimming in recreational waters: sea water, rivers, and lakes. In Britain it is common to swim at the seaside in the summer months, but given the current poor state of many beaches (in terms of their water quality) what, if any, are the consequences for health? Water quality is measured according to microbiological standards, in particular coliforms (relating to the bacterium *Escherichia coli*) and streptococci (other bacteria, of the genus *Streptococcus*). For example, European safety limits for faecal coliforms and streptococci are 100 per 100 ml.
  The classic studies of the health effects of bathing come from Cabelli and his colleagues (1979) in New York City, and later at other American beaches

Table 8.3 Gastroenteritis among bathers and non-bathers (unexposed group) on a sample of British beaches

Faecal streptococci exposure (per 100ml)	Number	Rate of gastroenteritis (per 100)
Unexposed	605	9.7
0–19	184	10.9
20–39	161	10.6
40–59	82	18.3[a]
60–79	57	28.1[a]
80+	23	30.4[a]

[a] Significantly different from unexposed (control) group
Source: Kay et al. (1994: 907)

(Cabelli et al., 1982). In the first study, rates of gastrointestinal symptoms were compared among groups of swimmers and non-swimmers, according to whether they attended a polluted or a non-polluted beach (where pollution levels were measured by counts of faecal coliforms). There was a statistically significant difference between the symptom rate for swimmers and non-swimmers at the polluted beach, but not at the unpolluted beach. Subsequent work on other beaches confirmed that swimming in faecally polluted waters carried a risk of gastroenteritis. British work has supported these findings. For example, those swimming from a beach in Kent, on the south coast of England, were significantly more likely than non-bathers to develop gastrointestinal symptoms and diarrhoea (Balarajan et al., 1991). In a very thorough randomized trial on British beaches Fleisher et al. (1993) recruited participants and allocated these randomly into swimming and non-swimming groups (see also Kay et al., 1994). Gastrointestinal symptoms were significantly higher among the bathers than in the non-bathing controls, even though exposure to levels of faecal streptococci was well below European Commission guidance levels. From table 8.3 it is clear that gastroenteritis appears when exposure to faecal streptococci is about 40 per 100 ml, well below the limit of 100 per 100 ml. The conclusion therefore seems to be that there is a significant burden of illness (albeit relatively mild and not lasting long) after contact with waters polluted by faeces, even though levels of faecal pollution are within current safety standards. In other parts of the world health risks are absent; for example, those swimming from beaches around Sydney, Australia, did not appear to be at risk of gastrointestinal illness (Hunter, 1997: 47).

Microbiological water quality, and the health risks, seem therefore to be spatially patterned. Care needs to be taken, however, in ascribing poor bathing-water quality exclusively to poor sewage disposal. As Jones and Obiri-Danso (1999) have demonstrated, indicator bacteria are not always derived from sewage effluent; pathogens such as faecal coliforms and faecal streptococci, as well as campylobacter, are as likely to emanate from non-point sources as from sewage outfalls. Birds and agricultural runoff, including that from grazing animals, are established sources of contamination near water bodies. Further – and on a particularly geographical note – there is considerable small-scale spatial, as well as temporal, variation in pollution indicators: "different orders of magnitude in the degree of faecal pollution may be encountered within a few metres of a single bathing water, or within a few hours at the same sampling point" (Obiri-Danso et al., 1999: 57).

Concluding Remarks

This chapter reveals sharp contrasts between the burden of disease in the developed and the developing worlds. In parts of Africa, South America, and the Indian sub-continent, issues of water quality revolve around the lack of safe drinking water (table 5.2 above, page 145) and the consequent outbreaks of severe infectious disease that are exacerbated by frequent natural disasters such as monsoon flooding. Vast numbers of people are at risk as a result. Although people in North America and Europe faced these risks in the nineteenth century, diseases such as cholera are now, in the developed world, "at a distance." Periodic flooding in the developed world may result in heavy financial losses to business and the individual, but the public health risks are tiny in comparison with those in the developing world. While those living in quite comfortable material circumstances may be at risk from infections due to *Campylobacter* and *Cryptosporidium*, the health impact is in terms of morbidity and not, in general, mortality. Similarly, we have seen that, in the developed world, public health concerns about water quality relate to elevated levels of some metals, or to water hardness, to possible exposure to organic and other contaminants from hazardous waste sites, and to bathing waters of doubtful quality. Here too, we are not dealing with thousands of deaths.

As in the previous chapter, the tension between "scientific" (positivist) investigations of disease and illness, and public perceptions of risk, has emerged as an issue. What is missing is a fuller account of the structural factors that determine whether or not safe, potable water is supplied to the community, whether in the developed or developing world. Whether this relates to the privatizing of water companies and their wish to maximize profits and perhaps to under-invest in improving water quality, or to the political and economic constraints in the third world that prevent millions from enjoying access to safe supplies of water, an investigation of these wider-scale structural determinants is surely merited.

FURTHER READING

By far the most comprehensive review of literature relating to water quality and health, on which I have drawn quite heavily, is that by Hunter (1997); this contains a lengthy bibliography that provides an excellent starting point for further research.

Cliff and Haggett (1988) consider Snow's work on cholera and apply a battery of spatial analytical methods to his classic Soho data. Mintz et al. (1998) paint a comprehensive global picture of contemporary and recent work on cholera. Learmonth (1988) has a chapter on schistosomiasis.

There is a huge literature on the health risks of hazardous waste disposal. A good overview is in the report written for the National Research Council (1991).

CHAPTER 9

HEALTH IMPACTS OF GLOBAL ENVIRONMENTAL CHANGE

The previous two chapters have focused on two broad areas of environmental hazard, those relating to poor air and water quality. These hazards, along with others (such as sources of radiation, contaminated food, and noise, for example), have been a common area of research for environmental epidemiologists over many years. As McMichael (1999: 106) has noted, such studies have tended to be of "specific, direct-acting, environmental hazards within a local setting." I want in this final chapter to turn attention to broader processes of environmental and ecological change, those whose consequences for health are in part direct (as are air and water pollution) but also indirect. In particular, many of these processes are global in their scale and while, as we shall see, some of the research they spawn has tended to be more local in scope, the health impacts are likely to be widespread. Put simply, we need to shift attention away from identifying and managing specific hazards, to looking at environment as "habitat" (McMichael, 1999).

There are two broad areas of environmental change I shall consider. First, I examine the health consequences of ozone losses from the upper atmosphere. Then I turn to the impacts of climate change; the extent to which climate warming is likely to have negative consequences for human health. The various health impacts are summarized in table 9.1, and the discussion in this chapter follows this framework. Since we have only relatively recently come to appreciate the nature and scale of these environmental changes there is still considerable uncertainty attached to the predicted health consequences. The "error bars" around some of the quantitative estimates of risk provided by researchers are therefore inevitably wide, and this needs to be borne in mind.

Stratospheric Ozone Depletion

Ultraviolet radiation from the sun is usually divided into three classes, depending on wavelength. Long-wave radiation (UVA) occupies the range 315–400 nanometres (nm), while UVB and UVC radiation occupy 280–315 nm

Table 9.1 Health impacts of ozone depletion and climate change

---

Ozone depletion
  Skin cancer
  Cataract
  Immunosuppression

Climate change
  Direct effects
  • thermal stress
  Indirect effects
  • vector-borne diseases
  • food poisoning
  • sea level rise
  • agriculture

Source: Based on Martens (1998: 5)

and 100–280 nm, respectively. The ozone layer in the upper atmosphere (stratosphere) ensures that none of the short-wave UVC radiation that can damage genetic material (DNA) reaches earth. It further absorbs much of the UVB radiation that also harms living things.

In 1985 research was published (Farman et al., 1985) that pointed to losses of ozone during spring in Antarctica, now popularly known as the "hole" in the ozone layer. But given the lack of human population in the region this is of less direct human concern than depletion at lower latitudes (Bentham, 1994). Specifically, as Madronich (1992) demonstrated, there have been losses of ozone amounting to about 6 percent at latitudes of 45°, in both hemispheres (figure 9.1), though this masks seasonal losses that are typically higher in winter months. Further, these are very broad regional averages, and losses may be much greater in certain areas.

Stratospheric ozone is depleted by chlorofluorocarbons (CFCs) and other so-called halocarbons, used in refrigeration, as aerosol propellants, and in solvents. On reaching the stratosphere they form very reactive chemicals, such as chlorine monoxide, which react with ozone to convert it back to oxygen. Despite agreements reached in Montreal in 1987 to reduce CFC usage (and subsequent toughening of this protocol in Copenhagen, in 1992, to phase out production and consumption of CFCs) it is anticipated that the atmospheric chlorine load will continue at a high level and that significant recovery of the ozone layer will not occur for at least another 50 years. Ironically, the production of low-level (tropospheric) ozone due to pollution, particularly in western Europe, helps to absorb some UVB radiation and diminish the effects of stratospheric ozone depletion.

What, then, of the consequences for health? Of the radiation that does get through to the earth's surface there is evidence that it can cause skin cancers and eye disease (cataracts). We need, then, to consider recent evidence on the incidence of such diseases. In so doing we shall, of course, need to give due

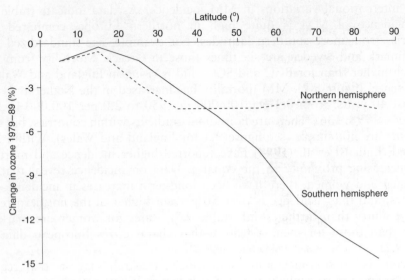

Figure 9.1 Changes in total ozone by latitude, northern and southern hemispheres, 1979–89 (Source: Bentham, G. (1994) Global environmental change and health, in Phillips, D.R. and Verhasselt, Y. (eds.) *Health and Development*, reproduced with kind permission of Routledge, London)

weight to the usual epidemiological issues of quantifying exposure and assessing the relationship between radiation dose and health response, as well as issues of potential confounding variables.

Ozone Depletion and Skin Cancer

There are three common forms of skin cancer, all of which are thought to be related to UV radiation (Department of the Environment, 1996). Two of these, basal cell carcinoma (BCC) and squamous cell carcinoma (SCC) are together referred to as *non-melanoma skin cancer* and account for about 90 percent of all skin cancers. The third, and most serious, is *malignant melanoma* (MM).

There is strong evidence that exposure to sunlight can cause non-melanoma skin cancer, but that this exposure occurred ten years or more prior to diagnosis. Cumulative lifetime exposure to sunlight seems to be a risk factor for SCC in particular, while adult risk of BCC is considered to be related to sun exposure in childhood and adolescence. Those working outdoors, in farming and fishing, are at risk, with cancers on exposed body areas, such as the head and neck. Incidence of non-melanoma is particularly high in Australia and New Zealand, where considerable proportions of the population have fair skin. The incidence is also higher among older people, but mortality rates from these cancers are generally low.

Malignant melanoma is much less common, but affects younger people as well as older age groups, and mortality rates are relatively high. What is known about international variations in MM incidence? As data indicate (table 9.2) the incidence of MM is quite high in northern Europe, compared with countries from more southern latitudes. For example, age-standardized rates in Denmark and Sweden are six times those in Greece. Mortality from MM is much higher than for BCC and SCC, and is rising in England and Wales as is incidence, (figure 9.2). MM mortality has increased in the Netherlands over the past 40 years, from 0.3 per 100,000 in 1950 to 2.0 per 100,000 in 1990 (Martens, 1998: 136). There are latitudinal gradients within countries, but these gradients are not simple (see figure 9.3 for England and Wales). Within New Zealand, Bulliard et al. (1994) have reported higher incidence and mortality with increasing proximity to the equator. Data on incidence reveal that, for both men and women, there have been long-term increases in incidence over 20 years, and that incidence is over 50 percent higher in the northern region of the country than further south (table 9.3). Rates for women are markedly higher than those for men, a feature that characterizes European data too (table 9.2).

In contrast to non-melanoma skin cancer, research suggests that regular, outdoor exposure to sunlight is not a risk factor. What matters more is intermittent, intense exposure of unacclimatized skin to sunlight. The increase in incidence and mortality is therefore thought to be due as much to behavioral changes in exposure, as illustrated by the rise in overseas vacations in hotter climates and the fashion for sun-tanned appearance. Evidence from the Department of the Environment (1996) suggests that participation in holidays abroad is socially patterned, with those in professional and managerial groups two or three times as likely as those in manual occupations to take foreign holidays. Similarly, participation in outdoor activities, such as golf and sailing is socially structured, with higher participation rates among higher-status occupations. Having said this, there is evidence that the use of suncreams is also higher in more affluent groups (see table 4.8, page 114).

Consequently, vacation and recreational behaviors, though not necessarily those involving skin protection, may go some way towards accounting for the social class gradient in MM, where the standardized mortality rate in professional groups is about 170, while for partly skilled and unskilled workers it is about 55. Gender differences in MM incidence are also of significance, with women being much more likely than men to have melanomas on their legs. Some public health experts therefore demand changes in health behaviors in order to reduce incidence. They point out that the contemporary preference for sun-tanned skin in many parts of the western world is socially constructed, in the sense that earlier centuries saw cultural value placed on white skin. They also indicate that people underestimate the risk of exposure to sunlight, in much the same way that other familiar risks controllable by the individual are underestimated, compared with those that are "managed" by the state and corporate interests (Bentham, 1993). Whether lower levels of use of (expensive) high-protection suncreams among the less affluent will attenuate the social class gradient in MM will only be apparent during the next two or three decades. It may be that such creams protect against sunburn but are less effective in protecting against skin cancer (Office for National Statistics, 1997: 61).

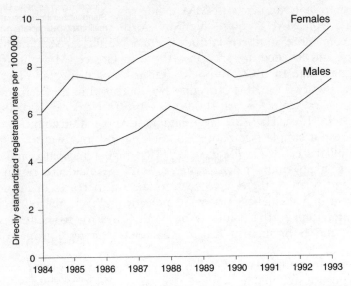

Figure 9.2   Incidence of malignant melanoma in England and Wales (1984–93)

Table 9.2 Malignant melanoma incidence rates in Europe (1996)

Country	All[a]	Females[a]	Males[a]
Denmark	12.9	14.1	11.3
Sweden	11.2	11.4	11.3
Netherlands	10.3	12.1	8.6
Austria	9.2	9.8	8.5
Ireland	8.0	10.2	5.9
Finland	7.5	6.7	8.7
France	7.1	7.6	6.5
United Kingdom	6.6	7.3	5.9
Germany	6.5	6.7	6.3
Belgium	6.0	6.3	5.6
Luxembourg	5.7	6.6	4.9
Italy	4.8	5.2	4.4
Spain	3.7	4.5	2.9
Portugal	3.1	3.6	2.4
Greece	1.9	1.9	1.8

[a] Age-standardized rates per 100,000

Source: International Agency for Research on Cancer (1999)

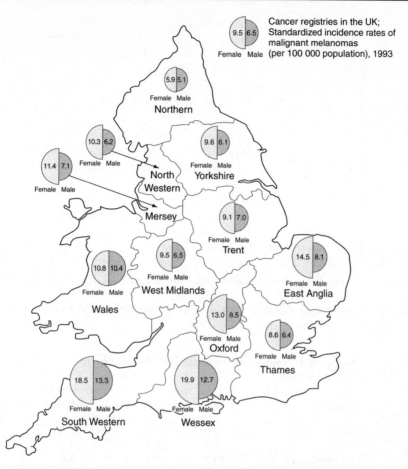

Figure 9.3 Standardized incidence rates of malignant melanoma in England and Wales, by region and sex, 1993

Table 9.3 Incidence of melanoma in New Zealand (1968–89)

Region	Average latitude	Men		Women	
		1968–73	1984–9	1968–73	1984–9
Northern	36°S	12.2	23.2	19.1	26.5
Midland	38°S	9.1	24.9	14.4	28.1
Central	41°S	7.6	21.9	11.4	25.8
Southern	44°S	6.6	18.3	12.1	24.4

Rates per 100,000, standardized to world population, non-Maori population only

Source: Bulliard et al. (1994: 236)

Let us look in more detail at the geographical incidence of MM in a particular part of the world, Scandinavia (Aase and Bentham, 1994; see also Aase and Bentham, 1996 and Bentham and Aase, 1996). Within countries such as Norway and Sweden there are regional variations, with some counties in the south having standardized registration ratios of over 125. But we have seen that there are, currently, social class gradients in MM, with those in professional groups more likely to have high rates of MM than those in manual occupations. To what extent, therefore, are regional variations in MM explained by variations in income (a proxy for lifestyles that might expose wealthier people to Mediterranean and Caribbean holiday destinations) rather than more local UV exposure?

Aase and Bentham set out to account for geographical variation in melanoma incidence (1970–9) using data on UVB exposure and socio-economic status. Given that monitored UVB levels are sparse, estimates of exposure were made for each county, based on latitude and cloudiness (itself estimated from other climatic data). Estimates of per capita income in each county were obtained. The authors also sought to test the hypothesis that ratios were higher in urban areas. A multiple regression analysis took MM incidence as the response variable and UVB exposure as the primary explanatory variable of interest; what effect does this have, when adjusting for other variables such as income?

A graph of (logarithmically transformed) MM ratios against estimated UVB exposure reveals a fairly clear association, but the relationship differs from country to country (figure 9.4). For example, at any given level of UVB exposure, Norwegian counties seem to have elevated ratios compared with Sweden and Finland. The multiple regression analysis confirms that the UVB index has a highly significant effect on MM. In other words, local domestic exposure to UVB carries a significant health risk.

Interestingly, Aase and Bentham go on to perform some simple "what if" analyses. In other words, they pose the question: how many extra cases of MM might we expect in each country if UVB exposure were to increase further? Results show that, for all five Nordic countries studied, there are approximately 2,800 cases per year. If UVB exposure were to increase by 10 percent, an entirely plausible scenario, this number would grow by nearly 400; if it were to increase by 50 percent, there would be well over 2,000 additional cases per annum.

These results give at least some credence to the importance of domestic rather than "exotic" exposure to UVB. For the authors, "it therefore seems that during the 1970s the dominant influence on variations in melanoma risk was not exposure to UVB on foreign holidays" (Aase and Bentham, 1994: 137). However, a later study, looking specifically at Norway, suggests that vacationing abroad is now beginning to have an impact on melanoma incidence (Bentham and Aase, 1996). Data show that while there has been a demonstrable effect of UVB levels on MM for more than 40 years, the impact of foreign holidays has only recently emerged as a significant predictor (table 9.4). We may need to wait for another 20 or 30 years to see what the real impact of recent and current vacation patterns has been.

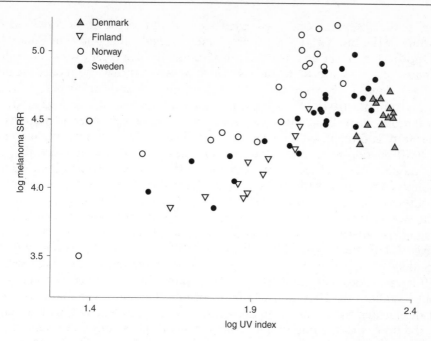

Figure 9.4   Relationship between incidence of melanoma and estimated UV radiation for counties in Scandinavia (Source: Aase, A. and Bentham, G. (1994) The geography of malignant melanoma in the Nordic countries: the implications of stratospheric ozone depletion, *Geografiska Annaler* 76B, 129–39), reproduced with kind permission of the Swedish Society for Anthropology and Geography.

Table 9.4   Regression analysis of relationship between incidence of malignant melanoma and UVB radiation and holidays abroad

| Time period | Regression coefficients | | R^2 (%) |
	UVB	Holidays abroad	
1955–9	0.177[a]	0.015	59
1960–4	0.212[a]	−0.024	79
1965–9	0.160[a]	0.039	89
1970–4	0.219[a]	−0.068	63
1975–9	0.274[a]	−0.030	88
1980–4	0.158[a]	0.084[a]	86
1985–9	0.184[a]	0.088[a]	91

[a] Significant at 0.05 level

Source: Bentham and Aase (1996: 1136)

Martens (1998) has conducted extensive research on the likely consequences of ozone depletion, with particular reference to the Netherlands and Australia. His approach is to set up a mathematical model of ozone depletion and to link estimates of UV skin dose to the incidence of skin cancer. The model includes the delay between exposure and tumour development, and various scenarios are explored on the basis of assumptions made about population size and age distribution, as well as rates of ozone loss. Modeling under conditions of uncertainty is fraught with danger, but nonetheless Martens envisages increases (by 2050) in non-melanoma of up to 60 cases per 100,000 in Australia, but quite modest increases (1 per 100,000) in MM incidence. Results indicate that although compliance with international agreements should see a reduction in chlorine concentrations and some stabilizing of ozone depletion in the early part of the twenty-first century, the lag between exposure and disease, coupled with an aging population, will lead to increases in all types of skin cancer until at least 2050. Slaper and his colleagues (1996) suggest that, even with the Copenhagen agreement to phase out ozone-depleting chemicals, there will by 2050 be an additional number of cases of skin cancer numbering (each year) about 33,000 in the USA, and 14,000 in Britain, Germany, Denmark, Belgium, and the Netherlands. This is a considerable health burden. But, as they point out, the impact of the international agreements to phase out CFCs should mitigate the effects on ozone depletion and therefore skin cancer; without these agreements the numbers would have been many times higher.

Other Health Impacts of Ozone Depletion

Although the bulk of the research effort on the health risks of UV radiation has been devoted to skin cancer, some researchers have pointed to possible impacts on eye disorders, particularly damage to the lens (increasing opaqueness or cataracts). There are three main types of cataract, and UVB radiation seems most heavily implicated in "cortical" cataract, in that it damages proteins in the lens.

A study of over 800 fishermen (known locally as "watermen") in Chesapeake Bay, USA, demonstrated a positive association between UVB exposure and cortical cataract; a doubling in UVB exposure over the lifetime is associated with a 60 percent increase in the risk of this cataract (Taylor, 1995). This study sought a rigorous measure of exposure, using detailed data on ambient UVB levels and personal exposure histories (including protection from the wearing of hats and sunglasses). The effect of UVB exposure was unaffected by adjustment for other possible risk factors. This is important, since merely demonstrating an increase in cataract with decreasing latitude, without adjusting for poor diet and poor material circumstances, as some studies have sought to do, is unhelpful. Other work (Cruickshanks et al., 1992) has confirmed these findings. Men living in Beaver Dam, Wisconsin, who were exposed to higher levels of UVB radiation, were 36 percent more likely to have severe cataract, even after adjusting for other risk factors. Research in Japan (Hayashi et al., 1998) examined the association between the prevalence of cataract in 47 districts (prefectures) and estimated UVB levels, adjusting for the proportion of the population that is over 75 years of age. There was a weakly significant relationship for women but not for men, though the study is hampered by poor measurement of exposure.

Armstrong (1994) has considered evidence which suggests that UVB radiation affects the development of immunity to natural infections. The longer-term consequences of this are hard to predict, but may lead to increases in infectious disease. Armstrong cautions that UVB could impair human response to immunization with live virus vaccines (such as that for measles), and that it can stimulate the replication of the AIDS virus (HIV) in the white blood cells ("T lymphocytes") that protect against viral infection. He also refers to convincing evidence that exposure to UVB radiation can reactivate previous infection with the herpes simplex virus (which causes cold sores).

On balance, there is plenty of evidence to suggest that UVB radiation causes skin cancer and some forms of cataract. It may also be implicated in other cancers, such as non-Hodgkin lymphoma; research by Langford et al. (1998) in Europe indicates that, after adjustment for socio-economic factors, estimated levels of UVB radiation predict variation in mortality from this disease. This work uses multi-level modeling, which we reviewed briefly in Chapter 3. It demonstrates a positive relationship between mortality and UVB in the United Kingdom and France, but in Italy the relationship is reversed.

From a health-promotion perspective the evidence suggests that exposure to the midday sun in summer months should be minimized and that wearing wide-brimmed hats is a simple but effective measure to counteract the effects of ozone layer depletion. A number of health campaigns along these lines are in place.

Global Climate Change

It is now well known that the additional burden of carbon dioxide, methane, and other gases, produced as the result of domestic, industrial, and agricultural activity, enhances the natural "greenhouse" effect, whereby such gases absorb and trap energy that is re-emitted by the earth's surface. Attempts to predict the magnitude of this global warming are fraught with difficulty and estimates must be treated very cautiously; but the UN's Intergovernmental Panel on Climate Change predicts an average increase globally of about 2°C by the year 2100, with a minimum of 1°C and a maximum of 3.5°C (Martens, 1998). Inevitably, it is difficult to appreciate what these numbers mean, in terms of daily experience, but an increase of, say, 2°C would mean that London would have temperatures similar to those in central France (Haggett, 1994). Global climate change models also predict an increase in precipitation, of between 7 and 15 percent. These estimates are more uncertain, but the best evidence suggests that increases in rainfall will be spatially uneven, with some areas becoming much wetter and others drier.

While global estimates of climate change are framed by uncertainty, and forecasts of place-to-place variations are also error-prone, we can speculate about, and also consider recent research evidence on, the possible or probable health consequences. Following McMichael and Haines (1997) and Martens (1998) we can think of these as *direct* effects – the impact of temperature increases on human physiology – and *indirect* effects, where the health impacts are mediated by the ways in which climate affects sea levels and ecosystem behaviour (table 9.1 above, page 235).

Direct Effects: Thermal Stress

Healthy people can cope well with changes in temperature; for example, leaving a warm house in the winter for a brisk walk on a cold evening will cause few problems for the fit person. But, outside a comfortable temperature range, as the body attempts to adjust to extreme cold and heat there can be risks to cardiovascular and respiratory health. These are likely to affect older people in particular. We consider here particularly the evidence concerning changes in mortality due to temperature increases.

Broadly speaking, there is a U-shaped relationship between mortality and environmental temperature. While mortality is high during extreme cold spells, it improves as the temperature rises; where average temperatures are below the comfort level, mortality improves by about 1 percent for every 1°C increase in average temperature. But as temperature rises above the comfort level there is an estimated 1.4 percent increase in mortality (Martens, 1998). In New York City, for example, deaths from all causes rise sharply if the temperature is higher than 33°C (91°F) (Kalkenstein, 1993). Looking specifically at cardiovascular mortality among those aged over 65 years, a unit increase in temperature in cold conditions reduces mortality by about 4 percent, while in very warm conditions it increases mortality by 1.6 percent. For deaths from respiratory causes there is a 3.8 percent reduction, and 10.4 percent increase, respectively. It appears from this that respiratory disease in particular is relatively sensitive to changes in temperature. These studies (see also Eurowinter Group, 1997) are drawn from cities in temperate climates, located in the developed world.

Martens (1998: 118–25) models the impact on mortality of possible global warming: results suggest that for places such as Singapore, where the climate is warm all year round, mortality will increase, but that for cities in colder climates (London, for example) modest increases in mortality in warmer months will be substantially offset by reductions in winter mortality. Evidence for cardiovascular mortality in selected countries (for those aged over 65 years) is shown in table 9.5. These scenarios are surrounded by a high degree of uncertainty and depend upon the ability of people to adapt physiologically to temperature change. Nonetheless, Martens' general conclusion is that global warming will probably reduce mortality, especially due to cardiovascular disease, because of warmer winters. Global climate change should therefore reduce excess winter mortality due to bronchitis, influenza, and heart disease.

Table 9.5 Estimated change in cardiovascular mortality
due to thermal stress (population aged over 65 years)

Country	Cold-related mortality change[a]	Warmth-related mortality change[a]
Singapore	0	43
Japan	−79	18
Netherlands	−181	19
UK	−250	10
USA	−184	32
Canada	−235	26
Spain	−129	33
Australia	−98	22

[a] Per 100,000 population

Source: Martens (1998: 123)

Indirect Effects

There may, of course, be other, more adverse, effects of global climate change. We should not neglect the interaction between global climate change and air pollution. Higher temperatures in summer months aid the photochemical reactions that produce low-level ozone (Chapter 7). In addition, research suggests that climate change will be associated with an increase in extreme weather events, and a rise in the incidence of storms and flooding. The flooding of low-lying coastal areas will bring with it direct loss of life, as well as potentially devastating impacts on infrastructure. We consider in this section the more indirect influences of such climate change. There is likely to be a considerable health burden arising from the indirect effects of global climate change (McMichael and Haines, 1997), because changes in temperature and rainfall will have major impacts on ecosystems. Ecosystem disturbance will manifest itself in terms of vector-borne disease, in infections leading to food poisoning, and in agricultural production, land use, and other change. We consider these in turn.

Impacts on infectious (especially insect-borne) disease

Research suggests that climate change will affect both the vectors and the infective agents that transmit infectious diseases such as malaria, dengue fever, and trypanosomiasis (sleeping sickness). Each is examined briefly.

Martens (1998) estimates that about 2.4 billion of the world's population is at risk from malaria, with between 300 and 500 million people suffering currently from the disease. Malaria is caused by the *Plasmodium* parasite (of which *P. vivax* and *P. falciparum* are most common in tropical areas and the latter is especially lethal), but the parasite is transmitted by species of the *Anopheles* mosquito, the saliva of which is injected into the human bloodstream. The ranges of the mosquito species are highly dependent on both temperature and rainfall, with an optimum temperature of about 20–25°C (68–77°F). Above this temperature mosquitoes will not survive. Similarly, while a minimum of 1.5 mm of rainfall a day is required, excess rainfall washes away mosquito larvae. But temperature also controls the survival of the parasite itself. Details of these climatic controls are given in Martens (1998).

Martens runs some global climate circulation models under different scenarios of climate change and finds that large parts of North America, Europe, Australia, and Asia may be at increased risk of malaria transmission; here, the *Anopheles* mosquitoes are already present but the *Plasmodium* parasite cannot currently survive because of low temperatures. Elsewhere, regional studies in Zimbabwe and the Andes of southern America suggest that the impact will be greatest in more upland areas within regions that are already malarial: malaria will spread into higher altitudes and therefore affect highland populations that are currently protected. Maps of the possible global impact, as well as scenarios in Zimbabwe, are shown in McMichael and Haines (1997). Martens' (1998) work suggests that the global population potentially at risk will probably rise from 2.4 billion to well over 3 billion, perhaps generating a further 220–480 million cases of the disease. The major burden will continue to be in tropical Africa.

A detailed case study of Rwanda is instructive (Loevinsohn, 1994). Rwanda (see box 6.1, page 172) had, in 1991, a population of about seven million, most of whom were living in rural highland areas. Loevinsohn shows that there have been long-term increases in temperature, though not in rainfall, since 1960, and a parallel increase in the recorded incidence of malaria, from about 35 per 1,000 people in 1982 to over 150 per 1,000 in 1990. Loevinsohn uses data from one health center, serving a population of 38,000, to examine smaller-scale variation. He finds that, between 1984 and 1987, incidence rose most sharply in high altitude areas and among children aged under two years. These are areas and a population group in which malaria had previously been rare. Monthly incidence between 1983 and 1990 is explained statistically by mean minimum temperature and rainfall, though the best-fitting model (explaining about 80 percent of the variation in incidence) is one relating incidence in any one month to climate data in the preceding two months. This is because the parasite takes 40–57 days to develop under the prevailing temperature conditions, and also because it takes time for the rainfall to collect in low-lying breeding sites. Loevinsohn attributes the increased incidence solely to these climatic influences; there were no other obvious explanations, such as a reduction in spraying programs to control mosquitoes or in-migration of non-immune people. The upsurge in malaria in Rwanda resulted from increases in temperature and rainfall.

In southern Asia malaria outbreaks seem to be associated with periods of particularly heavy rainfall (Lindsay and Birley, 1996). Historically, there has been a close correlation between deaths from malaria and the rainfall peaks that follow El Niño Southern Oscillation (ENSO) events (box 9.1; figure 9.5).

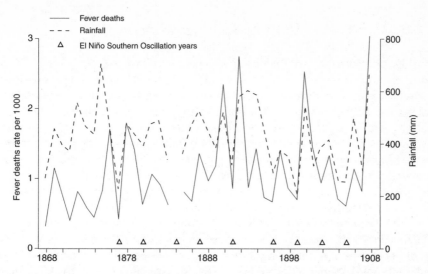

Figure 9.5 Relationship between deaths from malaria and rainfall in the Punjab, 1868–1908 (Source: Lindsay, S.W. and Birley, M.H. (1996) Climate change and malaria transmission, *Annals of Tropical Medicine and Parasitology*, 90, 573–88, reproduced with kind permission of the Liverpool School of Tropical Medicine)

Dengue fever is another vector-borne disease, the vector being the mosquito *Aedes aegypti*. It is a viral infection, the viruses (of which there are four serotypes) belonging to the genus *Flavivirus*. Infection varies from a relatively mild influenza-like illness to a severe form of haemorraghic disease (with bleeding from soft tissues) that is often fatal. The disease is widespread throughout much of Asia and Central and South America, with over 100 million cases reported each year. WHO indicates that the haemorraghic form affects children in particular and that mortality is about 5 percent, leading to 24,000 deaths per year. Rapid urbanization, population movement, the resistance of mosquitoes to insecticide, and the inadequate provision of piped water, are all implicated as factors in the increased incidence of dengue.

Temperature affects the distribution of the mosquitoes, the frequency with which they will bite, and also influences the incubation period of the virus; for example, at 27°C (81°F) the incubation period is 10 days, but this drops to seven days at 34°C (93°F) (Martens, 1998: 43). This shortening of the incubation period leads to an increase in the rate of transmission of the disease. Work by Jetten and Focks (1997) indicates that a warming of the global climate by 2°C would see transmission of dengue in some parts of southern Europe (Spain and Greece) and in the southern states of the USA. At some times of the year, for example late summer, current climates in North America would permit transmission of dengue. But, as with malaria, there is evidence that dengue is spreading into more highland areas in places in which it is already endemic, especially in Central and South America (McMichael, 1998: 23). For example, Mexico City, which is currently free from the disease because of its altitude, is sur-

BOX 9.1   *El Niño*

In order to understand the climate phenomenon known as El Niño, we need to know that, normally, off the west coast of South America cold air flows west to form the south-east trade winds, arriving in the west Pacific and then warming, rising, and at high altitudes flowing east to complete a circulation. But this pattern is disturbed every two to seven years. During the Christmas season (hence El Niño – "Christ child") the prevailing trade winds weaken and the warm surface waters that are normally driven west by the trade winds flow east towards South America. Strictly, El Niño refers to the warm ocean current flowing along the coast of Ecuador and Peru during December. The periodic, large-scale disturbance of ocean and air circulation is known as the El Niño Southern Oscillation (ENSO). This brings drought to countries bordering the Indian and Pacific Oceans and lengthy wet periods to Pacific regions. ENSO amplifies climate variability; in other words, periods of rainfall and drought are more intense than in unaffected areas. The opposite of El Niño is La Niña, characterized by unusually cool ocean temperatures in the eastern Pacific.

There is growing evidence that these events may lead to an increase in vector-borne disease. Nicholls (1993) suggests that the incidence of Australian encephalitis is associated with climatic extremes. There are outbreaks of the disease in south-east Australia between January and May, following heavy rainfall and flooding (accompanying a La Niña episode) which promotes the mosquito vector responsible for disease transmission. An unusually wet summer in 1974, again a La Niña event, led to an epidemic of fever in southern Africa, while in the previous year extreme monsoon conditions in India produced an epidemic of Japanese encephalitis in the northern Indian state of Uttar Pradesh. The death toll was 5,000, mostly children.

How does ENSO relate to global warming? Essentially, global temperature increases intensify or amplify the pattern of rainfall variability. An enhanced greenhouse effect may make ENSO events more frequent and more intense. As a result, the incidence of vector-borne diseases, where the vectors thrive on wet and warm conditions, is likely to increase. As Nicholls (1993: 1285) puts it, "changes in flood frequency, spatial extent, or severity, and changes in sea level, could result in centres of human activity and of vector activity coming closer, again increasing the risks of transmission."

rounded by lower-altitude, dengue-endemic areas and modest increases in temperature would permit the mosquitoes to survive at higher altitudes. Jetten and Focks (1997) suggest that Melbourne, Australia, is another city that might be affected.

Trypanosomiasis ("sleeping sickness") is carried by the tsetse fly (*Glossina morsitans*); the trypanosoma are parasitic protozoa whose hosts are wild and domestic animals. The disease, which is widespread in much of sub-Saharan

Africa, is fatal if left untreated. It tends to affect young and middle-aged adults, since they spend more time in fields and are therefore more likely to come into contact with the fly. As with other vector-borne diseases, authors have sought to relate data on the spatial distribution of the vector to environmental data, and then to assess the possible implications of climate change on its spread. For example, Rogers and Williams (1993) use GIS to model the distribution of the tsetse fly in east Africa. One climate variable, the maximum of the mean monthly temperature, is a highly significant predictor. Assuming mean increases in temperature of 1–3°C, highland areas of Zimbabwe become suitable for *Glossina*, indicating that, as with other vector-borne diseases, sleeping sickness may well spread to previously unaffected regions in tropical areas. There is a very clear role to be played by GIS and remote sensing, if informed by a good understanding of disease ecology, in the prediction of disease risk and in disease surveillance. Interestingly, Rogers and Williams (1993: 89–90) make the point that epidemics of sleeping sickness in Uganda over 100 years ago were quite possibly due to the movements of a colonial army and its porters. As they put it, "catastrophic epidemics are often associated with the arrival of a conquering army closely followed by the parasites to which the army has already adapted."

Rodents are implicated in some newly emerging, and re-emerging infectious diseases. For example, there was an outbreak of flu-like illness in the southwestern United States in 1993, with symptoms progressing to acute respiratory distress (Epstein, 1995). Investigations suggested that hantavirus was the culprit, a virus that is transmitted in the saliva and excreta of rodents (Duchin et al., 1994). But what caused the outbreak? Epstein suggests that heavy rainfall promoted the growth of vegetation and insects on which the rodents prey, causing a ten-fold increase in the rodent population over one year. He implies that climate change (El Niño events) may lie behind the outbreaks of rare infections.

Impacts on food poisoning

The fact that high summer temperatures in Britain are associated with food poisoning has led Bentham and Langford (1995) to model this association statistically, and then to use the statistical models as a basis for predicting the possible additional burden of illness, given assumptions about likely increases in temperature. This is very similar to the strategy adopted by Bentham and Aase in their work on skin cancer, described above (page 240).

Bentham and Langford take monthly data on the incidence of foodborne illness in England and Wales (1982–99), together with monthly mean temperatures. Reported cases of food poisoning (due largely to salmonella and campylobacter) are thought to be a considerable underestimate of true incidence, but these data are the best available. Data reveal (figure 9.6) a long-term upwards trend in notifications, as well as clear summer peaks. The secular trend is removed from the data, and analysis involves examining the relationship between the residuals (differences between observed notifications and trend) and temperature; the relationship to temperature in the month of notification and

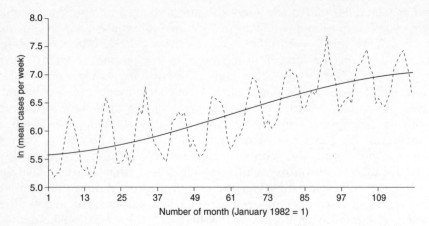

Figure 9.6 Reported weekly incidence of food poisoning in England and Wales, 1982–91 (Source: Bentham, G. and Langford, I.H. (1995) Climate change and the incidence of food poisoning in England and Wales, *International Journal of Biometeorology*, 39, 81–6, reproduced with kind permission of Springer-Verlag GmbH & Co. KG)

in the previous month is examined. Results (figure 9.7) show that there is a clear relationship between residuals and temperature during the previous month, but that this (almost linear) relationship is particularly striking at temperatures in excess of a threshold of 7.5°C (45.5°F).

Why should there be a significant link between incidence and previous month's temperature? Bentham and Langford (1995) suggest this is due to infection present in animals prior to their slaughter; the animals (chickens in particular) are then slaughtered, processed, and distributed. Their results imply that improvements in the storage and preparation of food by consumers are indeed required if food poisoning is to be reduced. Yet "blaming the victim" is not the entire panacea, since structural problems in the food industry need attention. Real improvements, according to Bentham and Langford, will come about by improving animal welfare and conditions in slaughterhouses (abattoirs) and by improving the preparation and storage of food by producers. Once more, we see that health outcomes are shaped not merely by individual human behavior, but by factors that are beyond the immediate control of the human actor.

What of the implications of global warming? Making some assumptions about likely temperature increases at particular times of the year, they predict increases of between 2.3 and 7.6 percent per month by 2010, but increases of up to 24 percent by 2050. These are simply rough guides, however, and if we assume that reported cases are a tiny fraction of true incidence we can suggest that real increases in cases by 2010 may be approaching 60,000, with perhaps 180,000 extra cases by 2050. The health and economic consequences (in terms of time off work, for example) of this are therefore not inconsequential.

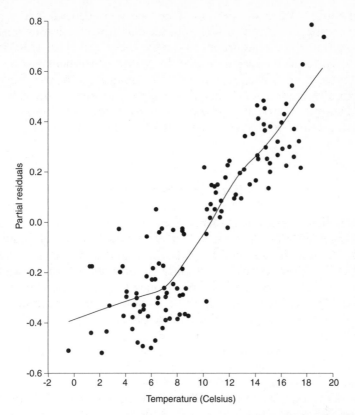

Figure 9.7 Relationship between regression residuals and temperature in previous month (Source: Bentham, G. and Langford, I.H. (1995) Climate change and the incidence of food poisoning in England and Wales, *International Journal of Bio-meteorology*, 39, 81–6, reproduced with kind permission of Springer-Verlag GmbH & Co. KG)

Other Health Effects of Climate Change

The Inter-governmental Panel on Climate Change (IPCC) has forecast sea-level rises of up to 40 cm (16 in.), perhaps more, by the turn of the next century. This will nearly double the number of people around the world who live in areas prone to flooding, to nearly 100 million (McMichael and Haines, 1997). The consequence of this will be population displacement, some of the health impacts of which we reviewed in Chapter 6, as well as outbreaks of disease such as cholera (Chapter 8). In August 2000 we witnessed precisely these outcomes in coastal areas of India. As with most impacts reviewed here, the burden falls on those least able to cope materially; environmental "inequity" will therefore be an even greater evil in the years ahead.

Climate change will also be likely to impact on agriculture. In particular, there will be regional variations in temperature change, in precipitation, and therefore in the ability of crops to germinate and grow. In general, experts predict a net negative impact on food production, although production in some temperate zones, such as the Canadian prairies, may actually increase (McMichael and Haines, 1997). Reduced agricultural yields will lead to increased malnutrition and hunger; we have already noted global inequalities in health outcomes and, more specifically, the impact on health in later life of low birth weight and poor nutritional status in early life. Estimates of the additional number of hungry people by the year 2060 range from 40–300 million (Parry and Rosenzweig, 1993).

Health Effects of Other Global Change

While the bulk of research evidence focuses on the health consequences of climate change, it is clear that the next 100 years will see other forms of global and environmental change that may also be significant. We can consider here processes of land use change, including deforestation and reforestation, as examples of other regional and global change processes. These are reviewed by Haggett (1994) and McMichael (1998). To some extent the diseases that are the focus of considerable public and epidemiological interest are those regarded as "emerging infectious diseases" (box 9.2).

The vector-borne Lyme disease may increase in incidence, either because of temperature increases or land use change. Lyme disease is caused by a spirochaete, *Borrelia burgdorferi*, which is transmitted by the tick vector *Ixodes ricinus* (Mawby and Lovett, 1998). This tick is a parasite on deer and mice. The disease was first recognized in the mid-1970s around the village of Old Lyme, Connecticut. Transmission of the spirochaete is influenced by temperature but also by land use change, since where abandoned farmland reverts to woodland the latter provides an ideal breeding-ground for the deer and mice hosts. As Mayer (2000) argues, deforestation of land on the outskirts of commuter suburbs in New England has created new habitats suitable for deer population. These ecological conditions are highly suitable for the transmission of the disease but, as Mayer suggests, the increasing incidence is very much a function of economic pressures to develop land.

During the 1990s about 12,500 cases were reported each year to the US Centers for Disease Control and Prevention (CDC). Most of the cases come from the north-east, but states such as Wisconsin and California also report large numbers of cases (see Kitron and Kazmierczak, 1997, for a study of the spatial distribution of the disease in Wisconsin). Those working outdoors, and more specifically in woodland and brush areas, as well as those participating in recreational activities such as camping and hunting, are at particular risk, though the burden of disease falls mostly on children and adults of middle age (peak incidences are in age groups 5–9 years, and 50–54 years). Given the reversion of farmland to woodland in parts of Europe it is no surprise that Lyme disease is increasing in incidence there too (Haggett, 1994). There is a strong geographical (west–east) gradient in risk, with fewer than one case per 100,000

BOX 9.2   *Emerging Infectious Diseases*

Until quite recently it was assumed, in the developed world, that infectious disease was largely a thing of the past and that the attention of public health professionals needed to be focused almost exclusively on chronic disease such as cancer and heart disease. The emergence of HIV and AIDS in the 1980s served as a warning that this conclusion was premature. More recently still, the emergence of various viral diseases (of which Lyme disease is perhaps the most well-known example) has meant that the study of "new" infections has become a considerable focus of research activity. Several examples are mentioned in this chapter, and these and others are considered by Mayer (2000).

One recent example, which gained publicity in 1999 in the USA, is West Nile Virus (WNV). This is a mosquito-borne disease that affects not only humans, but also horses and bird populations. It was first isolated in Africa in the 1930s. In its most severe form it causes encephalitis (inflammation of the brain), notably among older people. In August and September 1999 there were 62 such cases (and seven deaths) in New York City, nine of which were from a small area within northern Queen's Borough. A household survey, together with analysis of blood samples, suggested that about 2.6 percent of the population had been infected with the virus; many of these would have suffered few if any symptoms of the disease. Public health recommendations are to eliminate areas of standing and stagnant water, where the mosquitoes breed.

I think it is worth saying that the perceived impact of many of these diseases is greater than their real contribution to the burden of disease, especially in the developed world. Chronic disease will indeed continue to be the main problem here. Nonetheless, vigilance is required over new viral infections, and especially resurgent infections such as tuberculosis. In the developing world, worries over rare infections such as Ebola pale into insignificance when we consider the unimaginable burden of severe diarrhoeal disease, itself largely a function of poor infrastructure.

See page 254 below, and more particularly Farmer (1999: Chapter 2) for some critical reflections on the subject of "emerging" infections. Farmer's point is that many of these are not so new, but merely appear so to those living in comfortable conditions, compared with those in the developing world.

in the British Isles, yet 69 per 100,000 in southern Sweden and 130 per 100,000 reported in Austria (1995 data).

The clearance of forest and woodland in South America, and the consequent development of extensive monoculture has led to the emergence of new haemorrhagic fever viruses (arenaviruses). One example, from Venezuela, is Guaranito virus, confined to rural populations engaged in cattle-ranching. Like hantavirus, this has rodents as its host. Mortality rates are high. More broadly, the cutting of roads through rainforests encourages the development of water-collecting ponds at the roadside; these provide good breeding grounds for mosquitoes and the subsequent emergence of malaria. In Thailand, loss of forest has been associated with a rise in *Anopheles* mosquitoes. Using GIS, Gomes et al. (1998) have shown that in some districts the proportion of land area devoted to forest has fallen from 37 percent to 25 percent in ten years; forest has been replaced by commercial crops in plantations that provide suitable niches for mosquitoes. The new mosquito colonies "act as a reservoir of intense, unchecked malaria transmission, from which the workers who come to work during harvest periods can carry the parasite to their homes across Thailand" (Gomes et al., 1998: 94).

Concluding Remarks

This book has ended with a discussion of emerging infections, a discussion that arises out of a wider analysis of the health impacts of global environmental change. But while the last three chapters all dealt with some aspect of health and the physical environment, and earlier chapters considered the social, I think it is appropriate that this concluding section points to the links between the environmental and the social.

These links are summed up admirably by Epstein (1995: 170): "social, political and economic factors . . . are clearly integral to the condition and management of the environment, for therein lie the driving forces of global change . . . and *the inequitable distribution of exposures, vulnerabilities, and access to treatment*" (my emphasis). Mayer (2000) takes a broadly similar view, arguing that a "political ecology" framework (synthesizing traditional disease ecology with a political economic, or structuralist perspective) is required to understand the unintended health consequences of human-induced environmental change.

Getting to grips with an understanding of many of these emerging or resurgent infections requires a new form of geographical epidemiology, one that draws on an historical perspective but also on many of the social sciences – anthropology, sociology, economics, and politics. For Farmer (1999: 42–3), as with Urry's (2000) manifesto for a new sociology, we need to recognize that the appropriate units of analysis are not necessarily geographical ones such as nation-states; rather, they are flows of people, money, and viruses. We need a critical and a political epidemiology that examines how global financial and political institutions, institutionalized racism, and (neo)colonial history are implicated in disease aetiology. For writers such as Farmer, many diseases are not newly "emerging" or "re-emerging;" diseases such as tuberculosis may have declined in incidence within many developed countries, but for the poor in such countries, and many in the developing world, the notion of resurgence is a myth. In advancing the research agenda on links between global environmental change and health, it is essential that the "social" be retained. Geographies of health, at this global scale, must continue to draw on both environmental and social science.

FURTHER READING

The Department of the Environment (1996) produced a lengthy report into the effects, health and otherwise, of ozone depletion in the UK and is a useful source of further, more detailed evidence. Armstrong (1994) has a good review of research on ozone depletion and health.

A helpful introduction to the health effects of global climate change is provided in McMichael and Haines (1997), while Bentham (1994) reviews the literature on both the health impacts of ozone depletion and global warming. Martens (1998) offers a number of possible scenarios based on mathematical models. For a broad overview of research on the social science of climate change see the four-volume set: *Human Choice and Climate Change*, edited by S. Rayner and E.L. Malone (Battelle Press, Columbus, Ohio, 1998).

The collection of essays edited by Greenwood and de Cock (1998) is worth reading for various accounts of emerging infections and epidemics. See also Mayer's (2000) useful overview.

Finally, as I have indicated earlier, Farmer (1999) has written an important book which, fueled by an anthropological as well as epidemiological imagination, merits a reading by anyone concerned with health, space, and place.

APPENDIX: WEB-BASED RESOURCES FOR THE GEOGRAPHIES OF HEALTH

Introduction

The World Wide Web offers an unrivalled resource for the health researcher. Almost certainly, anyone picking up this book will be well used to "surfing the Net" in order to extract research and other material, and may well have visited some of the following sites in order to obtain information and perhaps data. I can guarantee that there are sites, especially those relating to particular countries, or diseases and disabilities, that are missing from the abbreviated list given below. But I think I can also guarantee that at least some of the sites listed provide adequate links to other, unlisted, sites that will provide additional, valuable material. In the last resort, using one of the "search engines" that are available, along with a judicious choice of keyword(s), will usually lead to something of value.

This Appendix is structured into two main sections. The first offers some sites that provide gateways into the research literature and good, general links, while the second lists some more specific sites of relevance to health geographers, including software.

A number of caveats are in order. First, I make no claims to completeness or comprehensiveness. Second, although I have visited all sites listed, the time lag between delivery of the manuscript to the publisher and its appearance as the book you are reading means that some sites may now be out of date; certainly, the information they contain will itself have been updated in many cases. You should therefore treat these sites as a set of signposts and routes rather than as a complete and accurate map. Third, although the pedigree of the sites listed here is quite impeccable, careful scrutiny of the "small print" will reveal that sources of data are of variable quality.

Please note that to save space I have not prefixed every site with http:// – this is implicit. But note for completeness that, for example, the WHO site that appears early in my list is available at:

http://www.who.int

General Sources

Accessing Relevant Literature

There are a number of on-line bibliographic search tools available, of which Medline is the most useful source. Bookmark the following site, which will be useful for obtaining abstracts of research papers:

www.ncbi.nlm.nih.gov/PubMed/

Some journals permit access on-line to full-length research papers. In some cases this involves an institution's Library subscribing to the relevant journals.

Major Health Sites

Cross-country comparisons

A good starting point is the World Health Organization's main site:

www.who.int

from which you can navigate to a wide range of sources and resources. In particular, their "Statistical Information System" has a wealth of data on health indicators, outcomes, diseases, health service provision, and so on:

www.who.int/whosis/

This site also has maps of global disease distribution.

There are similar sites for Regional Offices of WHO. For example, data on avoidable mortality in countries of central and eastern Europe (at the regional level) are available from the Danish (Copenhagen) office at:

www.who.dk/country

A huge amount of information is available for individual countries at UNICEF's web-site:

www.unicef.org

For example, data relating to its Progress of Nations program (1999), and its report on the State of the World's Children (1998), both of which contain numerous health indicators, may be found at:

www.unicef.org/pon99

and

www.unicef.org/sowc98

For further information about the Global Burden of Disease research outlined in Chapter 4, see:

www.hsph.harvard.edu/organizations/bdu/index.html

As indicated in Chapter 4, data for up to 15 major European cities are produced by the Mégapoles Project, administered by a health authority in London:

www.elcha.co.uk/holp/

USA

The best starting point is the Centers for Disease Control, based in Atlanta:

www.cdc.gov

and you can navigate from here to numerous relevant sites, including the National Center for Health Statistics; this collects, analyses and disseminates health statistics, including those on a state-by-state basis. The introductory page for NCHS is:

www.cdc.gov/nchswww/about/about.htm

For data on cancer incidence and mortality the American Cancer Society has a wealth of valuable material. Visit:

www.cancer.org/statistics

but see below for information relating to the Atlas of Cancer Mortality.
   A wealth of data on health services, including insurance coverage and service provision, may be obtained from the Statistical Abstract of the United States:

www.census.gov/statab/www/

The Morbidity and Mortality Weekly Report from the Centers for Disease Control has disease data at the state level as well as reports that may be downloaded:

www2.cdc.gov/mmwr/

UK

Although having only limited data disaggregated by geographical area there is useful material from the Office for National Statistics:

www.statistics.gov.uk

while the Department of Health site is also worth visiting for national data:

www.doh.gov.uk

This site contains controversial "performance monitoring" data for health authorities and hospital trusts in England and Wales. See:

www.doh.gov.uk/tables98/index.htm

Other countries

Inevitably, this is a tiny selection, but for those seeking further information relating to Canada, Australia, and New Zealand, respectively, see:

www.hc-sc.gc.ca/hppb/phdd/report/stat/eng/report.html
www.aihw.gov.au/publications/health/ah00.html
www.moh.govt.nz/moh.nsf/

The first of these is a Statistical Report for Canada, with vast amounts of health-related data by province. The second is a similarly voluminous report for Australia. The third relates more specifically to public health in New Zealand, and contains data on notifiable diseases for the 24 health districts.

Specific diseases

On-line searches will reveal information – and possibly data – on particular diseases. Research literature on particular diseases can be obtained via particular online databases, such as Medline, though some care is needed; for example, asking for all papers on "cancer" will be counter-productive! On the other hand, searching for papers on breast cancer survival in the USA will generate a more manageable list.

Again, WHO has numerous relevant sites, including those relating to tropical diseases such as malaria, dengue, dracunculiasis (guinea worm), and others. The schistosomiasis site, for example, is:

www.who.int/ctd/html/schisto.html

WHO also produces Epidemiological Fact Sheets, for example on HIV and AIDS, that contain data and maps for particular countries. Visit:

www.who.int/emc-hiv/fact_sheets/index.html

The Geography of Health

One extremely useful site, with numerous links (including those to many listed here) has been devised by health geographers at the University of Portsmouth, UK. See:

www.envf.port.ac.uk/geo/research/health/links

but see also the following:

http://geography.about.com/science/geography/msub19.htm

Margie Roswell at the University of Maryland has provided much useful information at the following site:

http://cgi.umbc.edu/~chpdm/healthgeo/

For a site relating to the historical work of Dr. John Snow (see Chapter 8), visit:

www.ph.ucla.edu/epi/snow.html

There is a growing number of lists containing news of events, meetings, publications, job opportunities, and so on. For example, Charles Croner (*ccroner@cdc.gov*) at the Office of Research and Methodology at the National Center for Health Statistics in the USA edits a regular newsletter relating to GIS and health: "Public Health GIS News and Information." This is oriented primarily towards North America but has much of general interest.

In Britain the Geography of Health Study Group of the Royal Geographical Society (with the Institute of British Geographers) has a regular newsletter. Visit the RGS web-site at:

www.rgs.org

and see the American equivalent (Association of American Geographers) at:

www.aag.org

Electronic Atlases and Associated Sources

Several sites contain maps or entire atlases of health-related maps, in some cases giving access to the data underlying such maps. A superb example is the Atlas of Cancer Mortality in the United States, 1950–94. See:

www.nci.nih.gov/atlas/

Examine the "HealthVis" system available at Pennsylvania State University; this allows the user to dynamically link different "views" of the data, such as a map with a scatterplot of rates and socio-demographic data:

www.geovista.psu.edu/NCHS/health.htm

The following site covers the use of remote sensing in monitoring human and animal health. It contains a bibliography and much other valuable material:

http://geo.arc.nasa.gov/sge/health/sensor/sensor.html

For the Atlas of Leading and Avoidable Causes of Death in Countries of Central and Eastern Europe, including data for 14 countries by administrative subdivision, see:

www.who.dk/country/readme.htm

Increasing amounts of data, and accompanying maps, relating to environmental quality are now available. The US Environmental Protection Agency website carries data on air quality and pollution, with some data provided in real-time, and collections of maps on a state-by-state basis. See:

www.epa.gov

Data are available here on hazardous waste ("Superfund") sites. See the maps at:

www.epa.gov/superfund/sites/npl/npl.htm

The Environment Agency in England and Wales also makes available reports and data:

www.environment-agency.gov.uk

Links to this and many other sites are available via Friends of the Earth:

www.foe.co.uk

Software

ESRI markets a number of GIS products, including ARC/INFO and the desktop package ArcView. Most academic institutions will have licences for these products, as will many health organizations. For some sample applications of GIS in the health field, look at the following page on ESRI's website:

www.esri.com/industries/health/health.htm

For information on the GIS package MapInfo, visit:

www.mapinfo.com

A variety of software products is becoming available for spatial analysis. Although devised for use in the study of crime, rather than health, CrimeStat may be used for performing some of the analyses (such as kernel estimation, and tests of clustering) discussed in Chapter 3. It is described in, and available free, from:

www.ojp.usdoj.gov/cmrc

Openshaw's Geographical Analysis Machine (GAM/K) is currently available to use without charge at:

www.ccg.leeds.ac.uk/smart/gam/gam.html

Relating to the Book

I have created, and will endeavor to maintain, a website for this book. It contains links to the sites listed above, and others. I shall try to keep these as up to date as I can, subject to other commitments. I will, of course, be grateful to readers who wish to supply suggestions for other sites that I can add as links. I cannot guarantee to do this, or to acknowledge who has provided the suggestions, but am happy to receive suggestions, modifications, or corrections nonetheless.

The address is:

www.lancs.ac.uk/users/IHR/gatrell/geohealth.htm

For further information about the publishers of the book, Blackwell, please see:

www.blackwellpublishers.co.uk

REFERENCES

Aase, A. and Bentham, G. (1994) The geography of malignant melanoma in the Nordic countries: The implications of stratospheric ozone depletion, *Geografiska Annaler* 76B, 129–39.

Aase, A. and Bentham, G. (1996) Gender, geography and socioeconomic status in the diffusion of malignant melanoma, *Social Science and Medicine*, 42, 1621–37.

Abdel-Wahab, M.F., Strickland, G.T., El-Sahly, A., El-Kady, N., Zakaria, S., and Ahmed, L. (1979) Changing pattern of schistosomiasis in Egypt 1935–1979, the *Lancet*, i, 242–4.

Adams, J. (1995) *Risk*, UCL Press, London.

Aggleton, P. (1990) *Health*, Routledge, London.

Akhtar, R. and Izhar, N. (1994) Spatial inequalities and historical evolution in health provision, in Phillips, D.R. and Verhasselt, Y. (eds.) *Health and Development*, Routledge, London.

Alderson, M. (1986) *Occupational Cancer*, Butterworth, London.

Alexander, F. (1993) Viruses, clusters and clustering of childhood leukaemia: A new perspective? *European Journal of Cancer*, 29A, 1424–43.

Altmann, P., Cunningham, J., Dhanesha, U., Ballard, M., Thompson, J., and Marsh, F. (1999) Disturbance of cerebral function in people exposed to drinking water contaminated with aluminium sulphate: Retrospective study of the Camelford water incident, *British Medical Journal*, 319, 807–11.

Anderson, H.R., de Leon, A.P., Bland, J.M., Bower, J.S., and Strachan, D.P. (1996) Air pollution and daily mortality in London, 1987–92, *British Medical Journal*, 312, 665–9.

Armstrong, B.K. (1994) Stratospheric ozone and health, *International Journal of Epidemiology*, 23, 873–85.

Arnold, D., ed. (1988) *Imperial Medicine and Indigeneous Societies*, Manchester University Press, Manchester.

Aronowitz, R.A. (1998) *Making Sense of Illness: Science, Society and Disease*, Cambridge University Press, Cambridge.

Ashton, J. ed. (1992), *Healthy Cities*, Open University Press.

Asthana, S. (1994) Primary health care and selective PHC: Community participation in health and development, in Phillips, D.R. and Verhasselt, Y. (eds.) *Health and Development*, Routledge, London.

Ayres, J.G. (1997) The health effects of air pollution in the United Kingdom, in Davison, G. and Hewitt, C.N. (eds.) *Air Pollution in the United Kingdom*, Royal Society of Chemistry, Cambridge.

Bailey, T.C. and Gatrell, A.C. (1995) *Interactive Spatial Data Analysis*, Addison Wesley Longman, Harlow.

Balarajan, R., Raleigh, V.S., Yuen, P., Wheeler, D., Machin, D., and Cartwright, R. (1991) Health risks associated with bathing in sea water, *British Medical Journal*, 303, 1444–5.

Banatavala, N., Roger, A.J., Denny, A., and Howarth, J.P. (1998) Mortality and morbidity among Rwandan refugees repatriated from Zaire, November, 1996, *Prehospital Disaster Medicine*, 13, 17–21.

Barker, D.J.P., ed. (1992) *Fetal and Infant Origins of Adult Disease*, BMJ Publishing Group, London.

Barker, D.J.P. (1994) *Mothers, Babies, and Disease in Later Life*, BMJ Publishing Group, London.

Bartley, M., Blane, D., and Davey Smith, G., eds. (1998) *The Sociology of Health Inequalities*, Blackwell, Oxford.

Beck, U. (1992) *Risk Society: Towards a New Modernity*, Sage Publications, London.

Benach, J. and Yasui, Y. (1999) Geographical patterns of excess mortality in Spain explained by two indices of deprivation, *Journal of Epidemiology and Community Health*, 53, 423–31.

Ben-Shlomo, Y. and Chaturvedi, N. (1995) Assessing equity in access to health care provision in the UK: Does where you live affect your chances of getting a coronary artery bypass graft? *Journal of Epidemiology and Community Health*, 49, 200–4.

Ben-Shlomo, Y. and Davey Smith, G. (1991) Deprivation in infancy or adult life: Which is more important for mortality risk? The *Lancet*, 337, 530–4.

Ben-Shlomo, Y., White, I., and Marmot, M. (1996) Does the variation in the socioeconomic characteristics of an area affect mortality? *British Medical Journal*, 312, 1013–4.

Bentham, G. (1986) Proximity to hospital and mortality from motor vehicle traffic accidents, *Social Science and Medicine*, 23, 1021–6.

Bentham, G. (1988) Migration and morbidity: Implications for geographic studies of disease, *Social Science and Medicine*, 26, 49–54.

Bentham, G. (1993) Depletion of the ozone layer: Consequences for non-infectious human disease, *Parasitology*, 106, 39–46.

Bentham, G. (1994) Global environmental change and health, in Phillips, D.R. and Verhasselt, Y. (eds.) *Health and Development*, Routledge, London.

Bentham, G. and Aase, A. (1996) Incidence of malignant melanoma of the skin in Norway, 1955–1989: Associations with solar ultraviolet radiation, income and holidays abroad, *International Journal of Epidemiology*, 25, 1132–8.

Bentham, G., Hinton, J., Haynes, R., Lovett, A. and Bestwick, C. (1995) Factors affecting non-response to cervical cytology screening in Norfolk, England, *Social Science and Medicine*, 40, 131–5.

Bentham, G. and Langford, I.H. (1995) Climate change and the incidence of food poisoning in England and Wales, *International Journal of Biometeorology*, 39, 81–6.

Benzeval, M. and Judge, K. (1996) Access to healthcare in England: Continuing inequalities in the distribution of general practitioners, *Journal of Public Health Medicine*, 18, 33–40.

Benzeval, M., Judge, K., and Whitehead, M. (1995) The role of the NHS, in Benzeval, M., Judge, K., and Whitehead, M. (eds.) *Tackling Inequalities in Health: An Agenda for Action*, The King's Fund, London.

Beral, V., Roman, E., and Bobrow, M., eds. (1993) *Childhood Cancer and Nuclear Installations*, BMJ Publishing, London.

Betemps, E.J. and Buncher, C.R. (1993) Birthplace as a risk factor in motor neurone disease and Parkinson's disease, *International Journal of Epidemiology*, 22, 898–904.

Bhandari, N.R., Syal, K., Kambo, I., Beohar, V., Sexena, N.C., Dabke, A.T., Agarwal, S.S. and Saxena, B.N. (1990) Pregnancy outcome in women exposed to toxic gas at Bhopal, *Indian Journal of Medical Research*, 92, 28–33.

Bhopal, R.S., Phillimore, P., Moffatt, S., and Foy, C. (1994) Is living near a coking works harmful to health? A study of industrial air pollution, *Journal of Epidemiology and Community Health*, 48, 237–47.

Bogard, W. (1989) *The Bhopal Tragedy,: Language, Logic, and Politics in the Production of a Hazard*, Westview Press, Boulder, Colorado.

Bollini, P. and Siem, H. (1995) No real progress towards equity: Health of migrants and ethnic minorities on the eve of the year 2000, *Social Science and Medicine*, 41, 819–28.

Borrell, C. and Arias, A. (1995) Socio-economic factors and mortality in urban settings: The case of Barcelona, Spain, *Journal of Epidemiology and Community Health*, 49, 460–5.

Borrell, C., Regidor, E., Arias, L.C., Navarro, P., Puigpinos, R., Dominguez, V., and Plasencia, A. (1999) Inequalities in mortality according to educational level in two large South European cities, *International Journal of Epidemiology*, 28, 58–63.

Boyle, P.J., Kudlac, H., and Williams, A.J. (1996) Geographical variation in the referral of patients with chronic end stage renal failure for renal replacement therapy, *Quarterly Journal of Medicine*, 89, 151–7.

Boyle, P., Halfacree, K., and Robinson, V. (1998) *Exploring Contemporary Migration*, Addison Wesley Longman, Harlow.

Boyle, P., Gatrell, A.C., and Duke-Williams, O. (1999) The effect on morbidity of variability in deprivation and population stability in England and Wales: An investigation at small-area level, *Social Science and Medicine*, 49, 791–9.

Bradley, D. (1998) The influence of local changes in the rise of infectious disease, in Greenwood, B. and de Cock, K. (eds.) *New and Resurgent Infections*, John Wiley, Chichester.

Brothwell, D. (1993) On biological exchanges between the Two Worlds, in Bray, W. (ed.) *The Meeting of the Two Worlds: Europe and the Americas 1492–1650*, Oxford University Press, Oxford.

Brown, M.P. (1997) *RePlacing Citizenship: AIDS Activism and Radical Democracy*, Guilford Press, London.

Brown, P. (1992) Popular epidemiology and toxic waste contamination: lay and professional ways of knowing, *Journal of Health and Social Behaviour*, 33, 267–81.

Brunekreef, B., Janssen, N.A.H., de Hartog, J., Harssema, H., Knape, M., and van Vliet, P. (1997) Air pollution from truck traffic and lung function in children living near motorways, *Epidemiology*, 8, 298–303.

Buchdahl, R., Parker, A., Stebbings, T., and Babiker, A. (1996) Association between air pollution and acute childhood wheezy episodes: Prospective epidemiological study, *British Medical Journal*, 312, 661–5.

Bulliard, J-L, Cox, B. and Elwood, J.M. (1994) Latitude gradients in melanoma incidence and mortality in the non-Maori population of New Zealand, *Cancer Causes and Control*, 5, 234–40.

Butchart, A. (1996) The industrial panopticon: Mining and the medical construction of migrant African labour in South Africa, *Social Science and Medicine*, 42, 185–97.

Butchart, A. (1998) *The Anatomy of Power: European Constructions of the African Body*, Zed Books, London.

Butler, R. and Parr, H., eds. (1999) *Mind and Body Spaces: Geographies of Illness, Impairment and Disability*, Routledge, London.

Cabelli, V.R., Dufour, A.P., Levin, M.A., McCabe, L.J., and Haberman, P.W. (1979) Relationship of microbial indicators to health effects at marine bathing beaches, *American Journal of Public Health*, 69, 690–6.

Cabelli, V.R., Dufour, A.P., McCabe, L.J., and Levin, M.A. (1982) Swimming associated gastroenteritis and water quality, *American Journal of Epidemiology*, 115, 606–16.

Cairney, J. and Ostbye, T. (1999) Time since immigration and excess body weight, *Canadian Journal of Public Health*, 90, 120–4.

Carballo, M. and Siem, H. (1996) Migration, migration policy and AIDS, in Haour-Knipe, M. and Rector, R. *Crossing Borders: Migration, Ethnicity and AIDS*, Taylor and Francis, London.

Carballo, M., Divino, J.J., and Zeric, D. (1998) Migration and health in the European Union, *Tropical Medicine and International Health*, 3, 936–44.

Carr-Hill, R., Rice, N., and Roland, M. (1996) Socioeconomic determinants of rates of consultation in general practice based on fourth national morbidity survey of general practices, *British Medical Journal*, 312, 1008–13.

Carr-Hill, R., Place, M., Posnett, J. (1997) Access and utilisation of health care services, in Ferguson, B., Sheldon, T. and Posnett, J. (eds.) *Concentration and Choice in Health-care*, Financial Times Healthcare, London.

Carrasquillo, O., Himmelstein, D.U., Woolhandler, S., and Bor, D.H. (1999) Trends in health insurance coverage, 1989–1997, *International Journal of Health Services*, 29, 467–83.

Chakraborti, A.K. and Saha, K.C. (1987) Arsenical dermatosis from tubewell water in West Bengal, *Indian Journal of Medical Research*, 85, 326–34.

Chapple, A. and Gatrell, A.C. (1998) Variations in use of cardiac services in England: perceptions of general practitioners, general physicians and cardiologists, *Journal of Health Services Research and Policy*, 3, 153–8.

Charlton, J. (1996) Which areas are healthiest? *Population Trends*, 83, 17–24.

Chen, C-J., Wu, M-M., Lee, S-S., Wang, J-D., Cheng, S-H., and Wu, H-Y. (1988) Artherogenicity and carcinogenicity of high arsenic artesian well water. Multiple risk factors and related malignant neoplasms of blackfoot disease, *Arteriosclerosis*, 8, 452–60.

Clarke, K., Howard, G.C.W., Elia, M.H., Hutcheon, A.W., Kaye, S.B., Windsor, P.M., and Yosef, H.M.A. (1995) Referral patterns within Scotland to specialist oncology centres for patients with testicular germ cell tumours, *British Journal of Cancer*, 72, 1300–2.

Cliff, A.D. and Haggett, P. (1988) *Atlas of Disease Distributions*, Blackwell, Oxford.

Cliff, A.D. and Smallman-Raynor, M.R. (1992) The AIDS pandemic: global geographical patterns and local spatial processes, *Geographical Journal*, 158, 182–98.

Cliff, A.D., Haggett, P., and Ord, J.K. (1986) *Spatial Aspects of Influenza Epidemics*, Pion, London.

Cliff, A.D., Haggett, P., and Smallman-Raynor, M. (1998) *Deciphering Global Epidemics: Analytical Approaches to the Disease Records of World Cities, 1888–1912*, Cambridge University Press, Cambridge.

Cliff, A.D., Haggett, P., and Smallman-Raynor, M. (2000) *Island Epidemics*, Oxford University Press, Oxford.

Cloke, P., Philo, C., and Sadler, D. (1991) *Approaching Human Geography*, Paul Chapman, London.

Cook-Mozaffari, P. (1996) Cancer and fluoridation, *Community Dental Health*, 13, 56–62.

Cooper, H., Arber, S., Fee, L., and Ginn, J. (1999) *The Influence of Social Support and Social Capital on Health: A Review and Analysis of British Data*, Health Education Authority, London.

Cornwell, J. (1984) *Hard-Earned Lives: Accounts of Health and Illness from East London*, Tavistock Publications, London.

Coulson, A. (1982) *Tanzania: A Political Economy*, Clarendon Press, Oxford.

Cousens, S.N., Linsell, L., Smith, P.G., Chandrakumar, M., Wilesmith, J.W., Knight, R.S.G., Zeidler, M., Stewart, G., and Will, R.G. (1999) Geographical distribution of variant CJD in the UK (excluding Northern Ireland), the *Lancet*, 353, 18–21.

Cox, B.D., Huppert, F.A., and Whichelow, M.J., eds. (1993) *The Health and Lifestyle Survey: Seven Years On*, Dartmouth, Aldershot.

Craddock, S. (1995) Sewers and scapegoats: Spatial metaphors of smallpox in nineteenth century San Francisco, *Social Science and Medicine*, 41, 957–68.

Craddock, S. (2000) Disease, social identity, and risk: Rethinking the geography of AIDS, *Transactions of the Institute of British Geographers*, 25, 153–68.

Cruickshanks, K.J., Klein, B.E., and Klein, R. (1992) Ultraviolet light exposure and lens opacities: The Beaver Dam Eye Study, *American Journal of Public Health*, 82, 1658–62.

Cullinan, P., Acquilla, S., and Dhara, V.R. (1997) Respiratory morbidity 10 years after the Union Carbide gas leak at Bhopal: A cross-sectional survey, *British Medical Journal*, 314, 338–42.

Curtin, P.D. (1989) *Death by Migration*, Cambridge University Press, Cambridge.

Curtis, S. and Jones, I.R. (1998) Is there a place for geography in the analysis of health inequality?, in Bartley, M., Blane, D. and Davey Smith, G. (eds.) *The Sociology of Health Inequalities*, Blackwell, Oxford.

Curtis, S. and Taket, A. (1996) *Health and Societies: Changing Perspectives*, Edward Arnold, London.

Curtis, S., Gesler, W., Smith, G., and Washburn, S. (2000) Approaches to sampling and case selection in qualitative research: examples in the geography of health, *Social Science and Medicine*, 50, 1001–14.

Darby, S., Whitley, E., Silcocks, P., Thakrar, B., Green, M., Lomas, P., Miles, J., Reeves, G., Fearn, T., and Doll, R. (1998) Risk of lung cancer associated with residential radon exposure in south-west England: A case-control study, *British Journal of Cancer*, 78, 394–408.

Davey, B. and Seale, C. (eds.) (1996) *Experiencing and Explaining Disease*, Open University Press, Buckingham.

Davey Smith, G., Neaton, J., Wentworth, D., Stamler, R., and Stamler, J. (1996a) Socio-economic differences in mortality risk among men screened for the multiple risk factor intervention trial: Part I – results for 300,685 white men, *American Journal of Public Health*, 86, 486–96.

Davey Smith, G., Wentworth, D., Neaton, J., Stamler, R., and Stamler, J. (1996b) Socio-economic differences in mortality risk among men screened for the multiple risk factor intervention trial: Part II – results for 20,224 black men, *American Journal of Public Health*, 86, 497–504.

Davidson, J. (2000) ". . . the world was getting smaller": Women, agoraphobia and bodily boundaries, *Area*, 32, 31–40.

Day, N. and many others (1999) Exposure to power-frequency magnetic fields and the risk of childhood cancer, the *Lancet*, 354, 1925–31.

Dear, M. and Wolch, J. (1987) *Landscapes of Despair: From Deinstitutionalization to Homelessness*, Polity Press, London.

Department of the Environment (1996) *The Potential Effects of Ozone Depletion in the United Kingdom*, Stationery Office, London.

DHSS (1980) *Inequalities in Health: Report of a Working Group*, Department of Health and Social Security, London.

Diggle, P.J. (1993) Point process modelling in environmental epidemiology, in Barnett, V. and Turkman, K.F. (eds.) *Statistics for the Environment*, John Wiley, Chichester.

Diggle, P.J. and Rowlingson, B.S. (1994) A conditional approach to point process modelling of elevated risk, *Journal of the Royal Statistical Society, Series A*, 157, 433–40.

Diggle, P.J., Gatrell, A.C., and Lovett, A.A. (1990) Modelling the prevalence of cancer of the larynx in part of Lancashire: A new methodology for spatial epidemiology, in Thomas, R.W. (ed.) *Spatial Epidemiology*, Pion, London.

Dissanayake, C.B. and Chandrajith, R.L.R. (1996) Iodine in the environment and endemic goitre in Sri Lanka, in Appleton, J.D., Fuge, R., and McCall, G.J.H. (eds.) *Environmental Geochemistry and Health*, the Geological Society, London.

Dobson, M.J. (1997) *Contours of Death and Disease in Early Modern England*, Cambridge University Press, Cambridge.

Dockery, D.W., Pope, C.A., Xu, X., Spengler, J.D., Ware, J.H., Fay, M.E., Ferris, B.G., and Speizer, F.E. (1993) An association between air pollution and mortality in six U.S. cities, *New England Journal of Medicine*, 329, 1753–9.

Dolk, H., Vrijheid, M., Armstrong, B., Abramsky, L., Bianchi, F., Garne, E., Nelen, V., Robert, E., Scott, J.E.S., Stone, D., and Tenconi, R. (1998) Risk of congenital anomalies near hazardous-waste landfill sites in Europe: The EUROHAZCON study, the *Lancet*, 352, 423–7.

Donald, P.R. (1998) The epidemiology of tuberculosis in South Africa, *Novartis Foundation Symposium*, 217, 24–41.

Dong, W., Ben-Shlomo, Y., Colhoun, H., and Chaturvedi, N. (1998) Gender differences in accessing cardiac surgery across England: A cross-sectional analysis of the Health Survey for England, *Social Science and Medicine*, 47, 1773–80.

Dorling, D. (1995) *A New Social Atlas of Britain*, John Wiley, Chichester.

Downing, A. and Sloggett, A. (1994) The use of the Longitudinal Study for environmental epidemiology, *Longitudinal Study Newsletter*, 11, 5–9.

Doyal, L. (1979) *The Political Economy of Health*, Pluto Press, London.

Drever, F. and Whitehead, M. (1995) Mortality in regions and local authority districts in the 1990s: Exploring the relationship with deprivation, *Population Trends*, 82, 19–26.

Drever, F. and Whitehead, M. (1997) *Health Inequalities: Decennial Supplement*, Office for National Statistics, London.

Du Plessis, J.B., van Rooyen, J.J., Naude, D.A., and van der Merwe, C.A. (1995) Water fluoridation in South Africa: Will it be effective? *Journal of the Dental Association of South Africa*, 50, 545–9.

Duchin, J.S., Koster, F.T., Peters, C.J., and others (1994) Hantavirus pulmonary syndrome: A clinical description of 17 patients with a newly recognised disease, *New England Journal of Medicine*, 330, 949–55.

Duncan, C., Jones, K., and Moon, G. (1996) Health-related behaviour in context: A multi-level modelling approach, *Social Science and Medicine*, 42, 817–30.

Dunn, C.E. and Kingham, S. (1996) Establishing links between air quality and health: Searching for the impossible? *Social Science and Medicine*, 42, 831–41.

Durkin, M.S., Davidson, L.L., Kuhn, L., O'Connor, P., and Barlow, B. (1994) Low-income neighborhoods and the risk of severe pediatric injury: A small-area analysis in northern Manhattan, *American Journal of Public Health*, 84, 587–92.

Dyck, I. (1995a) Hidden geographies: The changing lifeworlds of women with multiple sclerosis, *Social Science and Medicine*, 40, 307–20.

Dyck, I. (1995b) Putting chronic illness "in place": Women immigrants' accounts of their health care, *Geoforum*, 26, 247–60.

Dyck, I. (1998) Women with disabilities and everyday geographies: Home space and the contested body, in Kearns, R.A. and Gesler, W.M. (eds.) *Putting Health into Place: Landscape, Identity and Well-Being*, University of Syracuse Press, Syracuse, New York.

Dyck, I. (1999) Body troubles: Women, the workplace and negotiations of a disabled identity, in Butler, R. and Parr, H. (eds.) *Mind and Body Spaces: Geographies of Illness, Impairment and Disability*, Routledge, London.

Dyck, I., Lewis, N.D., and McLafferty, S., eds. (2001) *Geographies of Women's Health*, Routledge, London.

Elford, J. and Ben-Shlomo, Y. (1997) Geography and migration, in Kuh, D. and Ben-Shlomo, Y. (eds.) *A Life Course Approach to Chronic Disease Epidemiology*, Oxford University Press, Oxford.

Elliott, P.J., Hills, M., Beresford, J., Kleinschmidt, I., Jolley, D., Pattenden, S., Rodrigues, L., Westlake, A., and Rose, G. (1992a) Incidence of cancer of the larynx and lung near incinerators of waste solvents and oils in Great Britain, the *Lancet*, 339, 854–8.

Elliott, P., Cuzick, J., English, D., and Stern, R., eds. (1992b) *Geographical and Environmental Epidemiology*, Oxford University Press.

Elliott, S.J. and Gillie, J. (1998) Moving experiences: a qualitative analysis of health and migration, *Health and Place*, 4, 327–40.

Elliott, S.J., Taylor, S.M., Walter, S., Stieb, D., Frank, J., and Eyles, J. (1993) Modelling psychosocial effects of exposure to solid waste facilities, *Social Science and Medicine*, 37, 791–805.

Ellis, M. and Muschkin, C. (1996) Migration of persons with AIDS – a search for support from elderly parents? *Social Science and Medicine*, 43, 1109–18.

Epstein, P.R. (1995) Emerging diseases and ecosystem instability: New threats to public health, *American Journal of Public Health*, 85, 168–72.

Eurowinter Group (1997) Cold exposure and winter mortality from ischaemic heart disease, cerebrovascular disease, respiratory disease and all causes, in warm and cold regions of Europe, the *Lancet*, 349, 1341–6.

Eyles, J. and Donovan, J. (1986) Making sense of sickness and care: An ethnography of health in a West Midlands town, *Transactions of the Institute of British Geographers*, 11, 415–27.

Eyles, J., Taylor, S.M., Johnson, N., and Baxter, J. (1993) Worrying about waste: Living close to solid waste disposal facilities in southern Ontario, *Social Science and Medicine*, 37, 805–12.

Fang, J., Madhavan, S., Bosworth, W., and Alderman, M.H. (1998) Residential segregation and mortality in New York City, *Social Science and Medicine*, 47, 469–76.

Farman, J.C., Gardiner, B.G., and Shanklin, J.D. (1985) Large losses of total ozone in Antarctica reveal seasonal ClO_x/NO_x interaction, *Nature*, 315, 207–10.

Farmer, P. (1992) *AIDS and Accusation: Haiti and the Geography of Blame*, University of California Press, Berkeley.

Farmer, P. (1999) *Infections and Inequalities: The Modern Plagues*, University of California Press, Berkeley.

Ferguson, B., Sheldon, T.A., and Posnett, J., eds. (1997) *Concentration and Choice in Healthcare*, FT Healthcare, London.

Ferguson, D.E. (1979) The political economy of health and medicine in colonial Tanganyika, in Kaniki, M.H.Y (ed.) *Tanzania Under Colonial Rule*, Longman, London.

Fleisher, J.M., Jones, F., Kay, D., Stanwell-Smith, R., Wyer, M., and Morano, R. (1993) Water and non-water risk factors for gastroenteritis among bathers exposed to sewage-contaminated marine waters, *International Journal of Epidemiology*, 22, 698–708.

Flowerdew, R. and Martin, D., eds. (1997) *Methods in Human Geography*, Addison Wesley Longman, London.

Forster, D.P., Newens, A.J., Kay, D.W.K., and Edwardson, J.A. (1995) Risk factors in clinically diagnosed presenile dementia of the Alzheimer type: A case-control study in northern England, *Journal of Epidemiology and Community Health*, 49, 253–8.

Free, C., White, P., Shipman, C., and Dale, J. (1999) Access to and use of out-of-hours services by members of Vietnamese community groups in south London: A focus group study, *Family Practice*, 16, 369–74.

Frenkel, S. and Western, J. (1988) Pretext or prophylaxis? Racial segregation and malarial mosquitos in a British tropical colony: Sierra Leone, *Annals of the Association of American Geographers*, 78, 211–28.

Gardner, M.J. (1992) Childhood leukaemia around the Sellafield nuclear plant, in Elliott, P., Cuzick, J., English, D., and Stern, R. (eds.) *Geographical and Environmental Epidemiology*, Oxford University Press.

Gardner, M.J., Snee, M.P., Hall, A.J., Powell, C.A., Downes, S., and Terrell, J.D. (1990) Results of case-control study of leukaemia and lymphoma among young people near Sellafield nuclear plant in West Cumbria, *British Medical Journal*, 300, 423–9.

Garvin, T. (1995) "We're strong women": Building a community-university research partnership, *Geoforum*, 26, 273–86.

Gatrell, A.C. and Löytönen, M. eds. (1998) *GIS and Health*, Taylor and Francis, London.

Gatrell, A.C. and Senior, M.L. (1999) Health and healthcare applications, in Longley, P., Maguire, D., Goodchild, M., and Rhind, D. (eds.) *Geographical Information Systems: Principles and Applications*, John Wiley.

Gatrell, A.C., Bailey, T.C., Diggle, P.J., and Rowlingson, B.S. (1996) Spatial point pattern analysis and its application in geographical epidemiology, *Transactions of the Institute of British Geographers*, 21, 256–74.

Gatrell, A.C., Garnett, S., Rigby, J., Maddocks, A., and Kirwan, M. (1998) Uptake of screening for breast cancer in South Lancashire, *Public Health*, 112, 297–301.

Gatrell, A.C., Thomas, C., Bennett, S., Bostock, L., Popay, J., Williams, G., and Shah-tahmasebi, S. (2000) Understanding health inequalities: Locating people in geographical and social spaces, in Graham, H. (ed.) *Understanding Health Inequalities*, Open University Press.

Gatrell, A.C., Berridge, D., Bennett, S., Bostock, L., Thomas, C., Popay, J., and Williams, G. (2001) Local geographies of health inequalities, in Boyle, P. et al. (eds.) *The Geography of Health Inequalities in the Developed World*, Ashgate Press.

Gellert, G.A. (1993) International migration and control of communicable diseases, *Social Science and Medicine*, 37, 1489–99.

Gerhardt, U. (1989) *Ideas About Illness: An Intellectual and Political History of Medical Sociology*, Macmillan, London.

Geronimus, A.T., Bound, J., Waidmann, T.A., Hillemeier, M.M., and Burns, P.B. (1996) Excess mortality among blacks and whites in the United States, *New England Journal of Medicine*, 335 (21) 1551–8.

Geronimus, A.T., Bound, J., and Waidmann, T.A. (1999) Poverty, time, and place: Variation in excess mortality across selected US populations, 1980–1990, *Journal of Epidemiology and Community Health*, 53, 325–34.

Gesler, W. (1991) *The Cultural Geography of Health Care*, University of Pittsburgh Press, Pittsburgh.

Gesler, W. (1993) Therapeutic landscapes: Theory and a case study of Epidauros, Greece, *Environment and Planning D: Society and Space*, 11, 171–89.

Gesler, W.M. (1996) Lourdes: Healing in a place of pilgrimage, *Health and Place*, 2, 95–106.

Gesler, W. (1998) Bath's reputation as a healing place, in Kearns, R.A. and Gesler, W.M. (eds.) *Putting Health into Place: Landscape, Identity and Well-Being*, Syracuse University Press, Syracuse, New York.

Getis, A. and Ord, J.K. (1992) The analysis of spatial association by use of distance statistics, *Geographical Analysis*, 24, 189–206.

Getis, A. and Ord, J.K. (1999) Spatial modelling of disease dispersion using a local statistic: The case of AIDS, in Griffith, D.A., Amrhein, C.G., and Huriot, J-M. (eds.) *Econometric Advances in Spatial Modelling and Methodology: Essays in Honour of Jean Paelinck*, Kluwer.

Giggs, J.A., Ebdon, D.S., and Bourke, J.B. (1980) The epidemiology of primary acute pancreatitis in Greater Nottingham: 1969–1983, *Social Science and Medicine*, 26, 79–89.

Giggs, J.A., Bourke, J.B., and Katschinski, B. (1988) The epidemiology of primary acute pancreatitis in the Nottingham Defined population area, *Transactions of the Institute of British Geographers*, 5, 229–42.

Ginns, S.E. and Gatrell, A.C. (1996) Respiratory health effects of industrial air pollution: A study in east Lancashire, UK, *Journal of Epidemiology and Community Health*, 50, 631–5.

Gittelsohn, A. and Powe, N.R. (1995) Small area variations in health care delivery in Maryland, *Health Services Research*, 30, 295–317.

Gleeson, B. (1999) *Geographies of Disability*, Routledge, London.

Glenn, L.L., Beck, R.W., and Burkett, G.L. (1998) Effect of a transient, geographically localised economic recovery on community health and income studies with longitudinal household cohort interview method, *Journal of Epidemiology and Community Health*, 52, 749–57.

Gober, P. (1994) On geographic variation in abortion rates in the United States, *Annals of the Association of American Geographers*, 84, 230–50.

Goldberg, S.J., Lebowitz, M.D., Graver, E.J., and Hicks, S. (1990) An association of human congenital cardiac malformations and drinking water contaminants, *Journal of the American College of Cardiologists*, 16, 155–64.

Goldsmith, J.R. (1999) The residential radon-lung cancer association in US counties: A commentary, *Health Physics*, 76, 553–7.

Goma Epidemiology Group (1995) Public health impact of Rwandan refugee crisis: What happened in Goma, Zaire, in July 1994?, the *Lancet*, 345, 339–44.

Gomes, M., Linthicum, K., and Haile, M. (1998) Malaria: The role of agriculture in changing the epidemiology of malaria, in Greenwood, B. and de Cock, K. (eds.) *New and Resurgent Infections: Prediction, Detection and Management of Tomorrow's Epidemics*, John Wiley, Chichester.

Gorman, B.K. (1999) Racial and ethnic variation in low birthweight in the United States: Individual and contextual determinants, *Health and Place*, 5, 195–208.

Gould, P. and Wallace, R. (1994) Spatial structures and scientific paradoxes in the AIDS pandemic, *Geografiska Annaler*, 76B, 105–16.

Graham, R.P., Forrester, M.L., Wysong, J.A.,Rosenthal, T.C., and James, P.A. (1995) HIV/AIDS in the rural United States: Epidemiology and health services delivery, *Medical Care Research Review*, 52, 435–52.

Gray, A. ed. (1993) *World Health and Disease*, Open University Press, Buckingham.

Grbich, C. (1999) *Qualitative Research in Health: An Introduction*, Sage, London.

Greenwood, B. and de Cock, K., eds. (1998) *New and Resurgent Infections: Prediction, Detection and Management of Tomorrow's Epidemics*, John Wiley, Chichester.

Griffith, D.A., Doyle, P.G., Wheeler, D.C., and Johnson, D.L. (1998) A tale of two swaths: Urban childhood blood-lead levels across Syracuse, New York, *Annals of the Association of American Geographers*, 88, 640–65.

Griffith, J., Riggan, W.B., Duncan, R.C., and Pellom, A.C. (1989) Cancer mortality in US counties with hazardous waste sites and ground water pollution, *Archives of Environmental Health*, 44, 69–74.

Gunnell, D.J., Peters, T.J., Kammerling, R.M., and Brooks, J. (1995) Relation between parasuicide, suicide, psychiatric admissions, and socioeconomic deprivation, *British Medical Journal*, 311, 226–30.

Haggett, P. (1994) Geographical aspects of the emergence of infectious diseases, *Geografiska Annaler*, 76B, 91–104.

Haining, R. (1990) *Spatial Data Analysis in the Social and Environmental Sciences*, Cambridge University Press, Cambridge.

Harding, S. and Maxwell, R. (1997) Differences in mortality of migrants, in Drever, F. and Whitehead, M. (eds.) *Health Inequalities*, Office for National Statistics, London.

Harrison, R.M., Leung, P.L., Somervaille, L., Smith, R., and Gilman, E. (1999) Analysis of incidence of childhood cancer in the West Midlands of the United Kingdom in relation to proximity to main roads and petrol stations, *Occupational and Environmental Medicine*, 56, 744–80.

Hatch, M.C., Beyea, J., Nieves, J.W., and Susser, M. (1990) Cancer near the Three Mile Island nuclear plant: Radiation emissions, *American Journal of Epidemiology*, 132, 397–412.

Hatch, M.C., Wallenstein, S., Beyea, J., Nieves, J.W., and Susser, M. (1991) Cancer rates after the Three Mile Island nuclear accident and proximity of residence to the plant, *American Journal of Public Health*, 81, 719–24.

Hartog, R. (1993) Essential and non-essential drugs marketed by the 20 largest European pharmaceutical companies in developing countries, *Social Science and Medicine*, 37, 897–904.

Hartwig, G.W. and Patterson, K.D., eds. (1978) *Disease in African History: An Introductory Survey and Case Studies*, Durham, North Carolina.

Hayashi, L.C., Tamiya, N., and Yano, E. (1998) Correlation between UVB radiation and the proportion of cataract: An epidemiological study based on a nationwide patient survey in Japan, *Industrial Health*, 36, 354–60.

Haynes, R. (1993) Radon and lung cancer in Cornwall and Devon, *Environment and Planning A*, 25, 1361–66.

Haynes, R. and Bentham, G. (1982) The effects of accessibility on general practitioner consultations, out-patient attendances and inpatient admissions in Norfolk, England, *Social Science and Medicine*, 16, 561–9.

Heyman, B., ed. (1998) *Risk, Health and Health Care*, Arnold, London.

Hodgson, M.J. (1988) An hierarchical location-allocation model for primary health care delivery in a developing area, *Social Science and Medicine*, 26, 153–61.

Howe, G.M. (1972) *Man, Environment and Disease in Britain: A Medical Geography of Britain through the Ages*, David and Charles, Newton Abbott.

Hunter, P.R. (1997) *Waterborne Disease: Epidemiology and Ecology*, John Wiley, Chichester.

Independent Inquiry into Inequalities in Health (1998), the Stationery Office, London.

Jetten, T.H. and Focks, D.A. (1997) Potential changes in the distribution of dengue transmission under climate warming, *American Journal of Tropical Medicine and Hygiene*, 57, 285–97.

Jochelson, K., Mothibeli, M., and Leger, J-P. (1991) Human immunodeficiency virus and migrant labour in South Africa, *International Journal of Health Services*, 21, 157–73.

Johansson, L.M., Sundquist, J., Johansson, S-E., Bergman, B., Qvist, J., and Traskman-Bendz, L. (1997) Suicide among foreign-born minorities and native Swedes: An epidemiological follow-up study of a defined population, *Social Science and Medicine*, 44, 181–7.

Jones, A.P. and Bentham, G. (1995) Emergency medical service accessibility and outcome from road traffic accidents, *Public Health*, 109, 169–77.

Jones, C.M., Taylor, G.O., Whittle, J.G., Evans, D., and Trotter, D.P. (1997) Water fluoridation, tooth decay in 5 year olds, and social deprivation measured by the Jarman score: Analysis of data from British dental surveys, *British Medical Journal*, 315, 514–7.

Jones, K. (1991) Multilevel models for geographical research, *Concepts and Techniques in Modern Geography*, Environmental Publications, Norwich.

Jones, K. and Obiri-Danso, K. (1999) Non-compliance of beaches with the EU directives of bathing water quality: Evidence of non-point sources of pollution in Morecambe Bay, *Journal of Applied Microbiology*, Symposium Supplement, 85, 101S–7S.

Jones, K. and Duncan, C. (1995) Individuals and their ecologies: Analysing the geography of chronic illness within a multilevel modelling framework, *Health and Place*, 1, 27–40.

Jones, K. and Moon, G. (1987) *Health, Disease and Society: An Introduction to Medical Geography*, Routledge, London.

Joseph, A.E. and Hallman, B.C. (1998) Over the hill and far away: Distance as a barrier to the provision of assistance to elderly relatives, *Social Science and Medicine*, 46, 631–9.

Joseph, A.E. and Phillips, D.R. (1984) *Access and Utilization: Geographical Perspectives on Health Care Delivery*, Harper and Row, London.

Kalipeni, E. and Oppong, J. (1998) The refugee crisis in Africa and implications for health and disease: A political ecology approach, *Social Science and Medicine*, 46, 1637–53.

Kalkenstein, L.S. (1993) Health and climate change: direct impacts in cities, the *Lancet*, 342, 1397–9.

Kanaiaupuni, S.M. and Donato, K.M. (1999) Migradollars and mortality: The effects of migration on infant survival in Mexico, *Demography*, 36, 339–53.

Kaplan, B.A. (1988) Migration and disease, in Mascie-Taylor, C.G.N. and Lasker, G.W. (eds.) *Biological Aspects of Human Migration*, Cambridge University Press, Cambridge.

Kaplan, G.A. (1996) People and places: Contrasting perspectives on the association between social class and health, *International Journal of Health Services*, 26, 507–19.

Karasek, R. and Theorell, T. (1990) *Healthy Work: Stress, Productivity and the Reconstruction of Working Life*, Basic Books, New York.

Karasek, R., Baker, D., Marxer, F., Ahlbom, A., and Theorell, T. (1981) Job decision latitude, job demands and cardiovascular disease: A prospective study of Swedish men, *American Journal of Public Health*, 71, 694–705.

Kawachi, I., Colditz, G.A., Aschemo, A., Rimm, E.B., Giovannuci, E., Stampfer, M.J., and Willett, W.C. (1996) A prospective study of social networks in relation to total mortality and cardiovascular disease in men in the USA, *Journal of Epidemiology and Community Health*, 50, 245–51.

Kawachi, I., Kennedy, B.P., Lochner, K., and Prothrow-Stich, D. (1997) Social capital, income inequality, and mortality, *American Journal of Public Health*, 87, 1491–8.

Kay, D., Fleisher, J.M., Salmon, R.L., Jones, F., Wyer, M.D., Godfree, A.F., Zelenauch-Jacquotte, Z., and Shore, R. (1994) Predicting the likelihood of gastro-enteritis from sea bathing: Results from a randomised exposure, the *Lancet*, 344, 905–9.

Kearns, R.A. (1991) The place of health in the health of place: The case of the Hokianga Special Medical Area, *Social Science and Medicine*, 33, 519–30.

Kearns, R.A. and Gesler, W.M., eds. (1998) *Putting Health into Place: Landscape Identity and Wellbeing*, Syracuse University Press, New York.

Kendall, G.M., Miles, J.C.H., Cliff, J.D., Green, B.M.R., Muirhead, C.R., Dixon, D.W., Lomas, P.R., and Goodridge, S.M. (1994) *Exposure to radon in UK dwellings*, National Radiological Protection Board, Didcot, Oxfordshire.

Kim, Y-E., Gatrell, A.C., and Francis, B.J. (2000) The geography of survival after surgery for colorectal cancer in southern England, *Social Science and Medicine*, 50, 1099–1107.

King, R. (1995) Migrations, globalization and place, in Massey, D. and Jess, P. (eds.) *A Place in the World? Places, Culture and Globalization*, Open University Press.

Kington, R., Carlisle, D., McCaffrey, D., Myers, H., and Allen, W. (1998) Racial differences in functional status among elderly US migrants from the South, *Social Science and Medicine*, 47, 831–40.

Kinlen, L.J. and Petridou, E. (1995) Childhood leukaemia and rural population movements: Greece, Italy and other countries, *Cancer Causes and Control*, 6, 445–50.

Kinlen, L.J., Clarke, K., and Hudson, C. (1990) Evidence from population mixing in British New Towns 1946–85 of an infective basis for childhood leukaemia, the *Lancet*, 336, 577–82.

Kinlen, L.J., O'Brien, F., Clarke, K., Balkwill, A., and Matthews, F. (1993) Rural population mixing and childhood leukaemia: Effects of the North Sea oil industry in Scotland, including the area near Dounreay nuclear site, *British Medical Journal*, 306, 743–8.

Kitron, U. and Kazmierczak, J.J. (1997) Spatial analysis of the distribution of Lyme disease in Wisconsin, *American Journal of Epidemiology*, 145, 558–66.

Kloos, H., Higashi, G.I., Cattani, J.A., Schlinski, V.D., Mansour, N.S., and Murrell, K.D. (1983) Water contact behaviour and schistosomiasis in an Upper Egyptian village, *Social Science and Medicine*, 17, 545–62.

Kloos, H. et al. (1997) Spatial patterns of human water contact and *Schistosoma mansoni* transmission and infection in four rural areas in Machakos District, Kenya, *Social Science and Medicine*, 44, 949–68.

Knapp, K.K. and Hardwick, K. (2000) The availability and distribution of dentists in rural ZIP codes and primary health care professional shortage areas (PC-HPSA) ZIP codes: comparison with primary care providers, *Journal of Public Health Dentistry*, 60, 43–8.

Kuh, D. and Ben-Shlomo, Y., eds. (1997) *A Life Course Approach to Chronic Disease Epidemiology*, Oxford University Press, Oxford.

Kunst, A. (1997) *Cross-National Comparisons of Socio-Economic Differences in Mortality*, Department of Public Health, Erasmus University, Rotterdam, Netherlands.

Lagakos, S.W., Wessen, B.J., and Zelen, M. (1986) An analysis of contaminated well water and health effects in Woburn, Massachusetts, *Journal of the American Statistical Association*, 81, 583–96.

Langford, I.H. (1991) Childhood leukaemia mortality and population change in England and Wales 1969–73, *Social Science and Medicine*, 33, 435–40.

Langford, I.H. (1994) Using empirical Bayes estimates in the geographical analysis of disease risk, *Area*, 26, 142–9.

Langford, I.H., Bentham, G., and McDonald, A-L. (1998) Mortality from non-Hodgkin lymphoma and UV exposure in the European Community, *Health and Place*, 4, 355–64.

Lanska, D.J. (1997) Geographic distribution of stroke mortality among immigrants to the United States, *Stroke*, 28, 53–7.

Larson, E.H., Hart, L.G., and Rosenblatt, R.A. (1997) Is non-metropolitan residence a risk factor for poor birth outcome in the US, *Social Science and Medicine*, 45, 171–88.

Launoy, G., le Coutour, X., Gignoux, P., Pottier, D., and Dugleux, G. (1992) Influence of rural environment on diagnosis, treatment, and prognosis of colorectal cancer, *Journal of Epidemiology and Community Health*, 46, 365–7.

Learmonth, A. (1988) *Disease Ecology: An Introduction*, Blackwell, Oxford.

Leukaemia Research Fund (1990) *Leukaemia and Lymphoma: An Atlas of Distribution within Areas of England and Wales 1984–1988*, Leukaemia Research Fund Centre for Clinical Epidemiology, University of Leeds, Leeds.

Levine, R.J., Khan, M.R., D'Souza, S., and Nalin, D.R. (1976) Cholera transmission near a cholera hospital, the *Lancet*, ii, 84–86.

Lewis, N.D. and Kieffer, E. (1994) The health of women: Beyond maternal and child health, in Phillips, D.R. and Verhasselt, Y. (eds.) *Health and Development*, Routledge, London.

Lindsay, S.W. and Birley, M.H. (1996) Climate change and malaria transmission, *Annals of Tropical Medicine and Parasitology*, 90, 573–88.

Little, M.A. and Baker, P.T. (1988) Migration and adaptation, in Mascie-Taylor, C.G.N. and Lasker, G.W. (eds.) *Biological Aspects of Human Migration*, Cambridge University Press, Cambridge.

Litva, A. and Eyles, J. (1995) "Coming out": Exposing social theory in medical geography, *Health and Place*, 1, 5–14.

Loevinsohn, M. (1994) Climatic warming and increased malaria incidence in Rwanda, the *Lancet*, 343, 714–8.

Loffler, W. and Hafner, H. (1999) Ecological pattern of first admitted schizophrenics in two German cities over 25 years, *Social Science and Medicine*, 49, 93–108.

Longhurst, R. (1996) Refocusing groups: Pregnant women's geographical experiences of Hamilton, New Zealand/Aotearoa, *Area*, 28, 143–9.

Longley, P.A., Goodchild, M.F., Maguire, D.J., and Rhind, D.W., eds. (1999) *Geographical Information Systems*, John Wiley and Sons, Chichester.

López-Abente, G. (1998) Bayesian analysis of emerging neoplasms in Spain, in Gatrell, A.C. and Löytönen, M. (eds.) *GIS and Health*, Taylor and Francis, London.

Love, D. and Lindquist, P. (1995) The geographical accessibility of hospitals to the aged: a Geographic Information Systems analysis within Illinois, *Health Services Research*, 29, 627–51.

Lovett, A.A. and Gatrell, A.C. (1988) The geography of spina bifida in England and Wales, *Transactions of the Institute of British Geographers*, 13, 288–302.

Lubin, J.H. and Boice, J.D. (1997) Lung cancer risk from residential radon: Meta-analysis of eight epidemiologic studies, *Journal of the National Cancer Institute*, 89, 49–57.

Lubin, J.H., Boice, J.D., Edling, C., Hornung, R.W., Howe, G.R., and Kunz, E. (1995) Lung cancer in radon-exposed miners and estimation of risk from indoor exposure, *Journal of the National Cancer Institute*, 87, 817–27.

Lunt, N., Atkin, K., and Hirst, M. (1997) Staying single in the 1990s: Single-handed practitioners in the new National Health Service, *Social Science and Medicine*, 45, 341–9.

Macintyre, S. (1998) Social inequalities and health in the contemporary world: comparative overview, in Strickland, S.S. and Shetty, P.S. (eds.) *Human Biology and Social Inequality*, Cambridge University Press, Cambridge.

Macintyre, S., MacIver, S., and Sooman, A. (1993) Area, class and health: Should we be focusing on places or people? *Journal of Social Policy*, 22, 213–34.

Mackenzie, W.R., Hoxie, N.J., Proctor, M.E., Gradus, M.S., Blair, K.A., Peterson, D.E., Kazmierczak, J.J., Addiss, D.G., Fox, K.R., Rose, J.B., and Davis, J.P. (1994) A massive outbreak in Milwaukee of *Cryptosporidium* infection transmitted through the public water supply, *New England Journal of Medicine*, 331, 161–7.

MacKian, S. (2000) Contours of coping: Mapping the subject world of long-term illness, *Health and Place*, 6, 95–104.

Madge, C. (1997) Public parks and the geography of fear, *Tijdschrift voor Economische en Sociale Geografie*, 88, 237–50.

Madronich, S. (1992) Implications of recent total atmospheric ozone measurements for biologically active ultraviolet radiation reaching the earth's surface, *Geophysical Research Letters*, 19, 37–40.

Malone, J.B. et al. (1997) Geographic information systems and the distribution of *Schistosoma mansoni* in the Nile Delta, *Parasitology Today*, 13, 112–9.

Marks, S. and Andersson, N. (1990) The epidemiology and culture of violence, in Manganyi, N. and du Toit, A. (eds.) *Political Violence and the Struggle in South Africa*, Macmillan, London.

Marmot, M.G., Rose, G., Shipley, M., and Hamilton, P.J.S. (1978) Employment grade and coronary heart disease in British civil servants, *Journal of Epidemiology and Community Health*, 3, 244–9.

Marmot, M.G., Davey Smith, G., Stansfield, S., and others (1991) Health inequalities among British civil servants: The Whitehall II study, the *Lancet*, 337, 1387–93.

Marmot, M.G. and Wilkinson, R.G., eds. (1999) *Social Determinants of Health*, Oxford University Press, Oxford.

Marshall, R.J. (1991) Mapping disease and mortality rates using empirical Bayes estimators, *Applied Statistics*, 40, 283–94.

Martens, P. (1998) *Health and Climate Change: Modelling the Impacts of Global Warming and Ozone Depletion*, Earthscan Publications, London.

Martin, D.J. (1996) *Geographic Information Systems: Socioeconomic Applications*, Routledge, London, second edition.

Martyn, C.N., Barker, D.J.P., and Osmond, C. (1989) Geographical relationship between Alzheimer's disease and aluminium in drinking water, the *Lancet*, i, 59–62.

Mascie-Taylor, C.G.N. and Lasker, G.W., eds. (1988) *Biological Aspects of Human Migration*, Cambridge University Press, Cambridge.

Mawby, T.V. and Lovett, A.A. (1998) The public health risks of Lyme disease in Breckland, U.K.: An investigation of environmental and social factors, *Social Science and Medicine*, 46, 719–27.

Mayer, J.D. (2000) Geography, ecology and emerging infectious diseases, *Social Science and Medicine*, 50, 937–52.

McCarthy, M. (1999) Transport and health, in Marmot, M. and Wilkinson, R.G. (eds.) *Social Determinants of Health*, Oxford University Press, Oxford.

McConway, K., ed. (1994) *Studying Health and Disease*, Open University Press, Buckingham.

McLafferty, S. (1982) Neighborhood characteristics and hospital closure, *Social Science and Medicine*, 16, 1667–74.

McLafferty, S. and Tempalski, B. (1995) Restructuring and women's reproductive health: Implications for low birthweight in New York City, *Geoforum*, 26, 309–23.

McLoone, P. and Boddy, F.A. (1994) Deprivation and mortality in Scotland, 1981 and 1991, *British Medical Journal*, 309, 1465–70.

McMichael, A.J. (1998) The influence of historical and global changes upon the patterns of infectious diseases, in Greenwood, B. and de Cock, K. (eds.) *New and Resurgent Infections: Prediction, Detection and Management of Tomorrow's Epidemics*, John Wiley, Chichester.

McMichael, A.J. (1999) Widening the frame: Environment as "habitat", not mere repository of "hazard", in Jedrychowski, W., Vena, J., and Maugeri, U. (eds.) *Challenges to Epidemiology in Changing Europe*, Polish Society for Environmental Epidemiology, Krakow, Poland.

McMichael, A.J. and Haines, A. (1997) Global climate change: The potential effects on health, *British Medical Journal*, 315, 805–9.

McNally, N.J., Williams, H.C., Phillips, D.R., Smallman-Raynor, M., Lewis, S., Venn, A., and Britton, J. (1998) Atopic eczema and domestic water hardness, the *Lancet*, 352, 527–31.

McNeil, D. (1996) *Epidemiologic Research Methods*, John Wiley, Chichester.

Melrose, D. (1982) *Bitter Pills: Medicines and the Third World Poor*, Oxfam Press, Oxford.

Mintz, E.D., Tauxe, R.V., and Levine, M.M. (1998) The global resurgence of cholera, in Noah, N. and O'Mahony, M. (eds.) *Communicable Disease Epidemiology and Control*, John Wiley, Chichester.

Miringoff, M. and Miringoff, M-L. (1999) *The Social Health of the Nation: How America is Really Doing*, Oxford University Press, Oxford.

Moffatt, S., Phillimore, P., Bhopal, R., and Foy, C. (1995) "If this is what it's doing to our washing, what is it doing to our lungs?" Industrial air pollution and public understanding in North-East England, *Social Science and Medicine*, 41, 883–91.

Mohan, J. (1998) Explaining geographies of health: A critique, *Health and Place*, 4, 113–24.

Monmonier, M. (1996) *How to Lie with Maps*, University of Chicago Press, Chicago, second edition.

Moon, G. and Brown, T. (1998) Place, space, and health service reform, in Kearns, R.A. and Gesler, W.M. (eds.) *Putting Health into Place: Landscape Identity and Wellbeing*, Syracuse University Press, New York.

Moore, A.J. (1995) Deprivation payments in general practice: Some spatial issues in resource allocation in the UK, *Health and Place*, 1, 121–5.

Morris, J.A., Butler, R., Flowerdew, R., and Gatrell, A.C. (1993) Retinoblastoma in children of former residents of Seascale, *British Medical Journal*, 306, 650.

Morris, R. and Carstairs, V. (1991) Which deprivation? A comparison of selected deprivation indexes, *Journal of Public Health Medicine*, 13, 318–26.

Mukerjee, M. (1995) Persistently toxic: The Union Carbide accident in Bhopal continues to harm, *Scientific American*, 252, June 8, 10.

Muller, I., Smith, T., Mellor, S., Rare, L., and Genton, B. (1998) The effect of distance from home on attendance at a small rural health centre in Papua New Guinea, *International Journal of Epidemiology*, 27, 878–84.

Muntaner, C. and Lynch, J. (1999) Income inequality, social cohesion, and class relations: A critique of Wilkinson's neo-Durkheimian research program, *International Journal of Health Services*, 29, 59–81.

Murray, C.J.L. and Lopez, A.D., eds. (1996) *Quantifying Global Burden Health Risks: The Burden of Disease Attributable to Selected Risk Factors*, Harvard University Press, Cambridge, Mass.

Murray, C.J.L. and Lopez, A.D. (1997) Global mortality, disability, and the contribution of risk factors: Global Burden of Disease study, the *Lancet*, 349, 1436–42.

National Research Council (1991) *Environmental Epidemiology: Public Health and Hazardous Wastes*, National Academy Press, Washington.

Nazroo, J. (1998) Genetic, cultural or socio-economic vulnerability? Explaining ethnic inequalities in health, in Bartley, M., Blane, D., and Davey Smith, G. (eds.) *The Sociology of Health Inequalities*, Blackwell, Oxford.

Neilson, S., Robinson, I., and Rose, F.C. (1996) Ecological correlates of motor neuron disease mortality: Hypothesis concerning an epidemiological association with radon gas and gamma exposure, *Journal of Neurology*, 243, 329–36.

Nemet, G.F. and Bailey, A.J. (2000) Distance and health care utilization among the rural elderly, *Social Science and Medicine*, 50, 1197–1208.

Neumann, C.M., Forman, D.L., and Rothlein, J.E. (1998) Hazard screening of chemical releases and environmental equity analysis of populations proximate to toxic release inventory facilities in Oregon, *Environmental Health Perspectives*, 106, 217–26.

Nicholls, N. (1993) El Niño-Southern Oscillation and vector-borne disease, the *Lancet*, 342, 1284–5.

Obiri-Danso, K., Jones, K., and Paul, N. (1999) The effect of time of sampling on the compliance of bathing water in NW England with the EU Directive on bathing water quality, *Journal of Coastal Conservation*, 5, 51–8.

O'Campo, P., Xue, X., Wang, M-C., and Caughy, M. (1997) Neighborhood risk factors for low birthweight in Baltimore: A multilevel analysis, *American Journal of Public Health*, 87, 1113–8.

Office for National Statistics (1997) *Health in England 1996: What People Know, What People Think, What People Do*, Stationery Office, London.

Openshaw, S. (1983) The Modifiable Areal Unit Problem, *Concepts and Techniques in Modern Geography*, Environmental Publications, Norwich.

Openshaw, S., Craft, A.W., Charlton, M., and Birch, J.M. (1988) Investigation of leukaemia clusters by use of a geographical analysis machine, the *Lancet*, i, 272–3.

Packard, R.M. (1989) *White Plague, Black Labor: Tuberculosis and the Political Economy of Health and Disease in South Africa*, University of California Press, Berkeley, California.

Paigen, B., Goldman, L.R., Magnant, M.M., Highland, J.H., and Steegmann, A.T. (1987) Growth of children living near the hazardous waste site, Love Canal, *Human Biology*, 59, 489–508.

Pain, R.H. (1997) Social geographies of women's fear of crime, *Transactions of the Institute of British Geographers*, 22, 231–44.

Paneth, N., Vinten-Johansen, P., Brody, H., and Rip, M. (1998) A rivalry of foulness: Official and unofficial investigations of the London cholera epidemic of 1854, *American Journal of Public Health*, 88, 1545–53.

Parr, H. (1998) The politics of methodology in "post-medical geography": Mental health research and the interview, *Health and Place*, 4, 341–54.

Parry, M.L. and Rosenzweig, C. (1993) Food supply and the risk of hunger, the *Lancet*, 342, 1345–7.

Pawson, E. and Banks, G. (1993) Rape and fear in a New Zealand city, *Area*, 25, 55–63.

Perry, B. and Gesler, W. (2000) Physical access to primary health care in Andean Bolivia, *Social Science and Medicine*, 50, 1177–88.

Pershagen, G., Liang, Z-H., Hrubec, Z., Svensson, C., and Boice, J.D. (1992) Residential radon exposure and lung cancer in Swedish women, *Health Physics*, 63, 179–86.

Petersen, A. and Lupton, D. (1996) *The New Public Health: Health and Self in the Age of Risk*, Sage, London.

Phillimore, P. (1993) How do places shape health? Rethinking locality and lifestyle in North-East England, in Platt, S., Thomas, H., Scott, S., and Williams, G. (eds.) *Locating Health: Sociological and Historical Explanations*, Avebury, Aldershot.

Phillimore, P. and Moffatt, S. (1994) Discounted knowledge: Local experience, environmental pollution and health, in Popay, J. and Williams, G. (eds.) *Researching the People's Health*, Routledge, London.

Phillips, D.R. (1990) *Health and Health Care in the Third World*, Longman, Harlow.

Phillips, D.R. (1991) Problems and potential of researching epidemiological transition: Examples from southeast Asia, *Social Science and Medicine*, 33, 395–404.

Phillips, D.R. (1994) Epidemiological transition: implications for health and health care provision, *Geografiska Annaler*, 76B, 71–89.

Phillips, D.R. and Verhasselt, Y., eds. (1994) *Health and Development*, Routledge, London.

Philo, C. (1989) "Enough to drive one mad": The organization of space in 19th-century lunatic asylums, in Wolch, J. and Dear, M. (eds.) *The Power of Geography: How Territory Shapes Social Life*, Unwin Hyman, Boston.

Philo, C. (1996) Staying in? Invited comments on "Coming out: exposing social theory in medical geography", *Health and Place*, 2, 35–40.

Philo, C. (1997) Across the water: Reviewing geographical studies of asylums and other mental health facilities, *Health and Place*, 3, 73–89.

Pocock, S.J., Shaper, A.G., Cook, D.G., Packham, R.F., Lacey, R.F., Powell, P., and Russell, P.F. (1980) British Regional Heart Study: Geographic variations in cardiovascular mortality, and the role of water quality, *British Medical Journal*, 280, 1243–8.

Pocock, S.J., Cook, D.G., and Shaper, A.G. (1982) Analysing geographic variation in cardiovascular mortality: Methods and results, *Journal of the Royal Statistical Society, Series A*, 145, 313–41.

Poland, B.D. (1998) Smoking, stigma, and the purification of public space, in Kearns, R.A. and Gesler, W.M. (eds.) *Putting Health into Place: Landscape Identity and Wellbeing*, Syracuse University Press, New York.

Porsch-Oezcueruemez, M. and others (1999) Prevalence of risk factors of coronary heart disease in Turks living in Germany: The Giessen study, *Atherosclerosis*, 144, 185–98.

Powell, D. and Leiss, W. (1997) *Mad Cows and Mother's Milk: The Perils of Poor Risk Communication*, McGill-Queen's University Press, Montreal.

Power, C., Manor, O., and Matthews, S. (1999) The duration and timing of exposure: effects of socio-economic environment on adult health, *American Journal of Public Health*, 89, 1059–65.

Price, P. (1994) Maternal and child health care strategies, in Phillips D.R. and Verhasselt, Y. (eds.) *Health and Development*, Routledge, London.

Pringle, D.G. (1998) Hypothesised foetal and early life influences on adult heart disease and mortality: An ecological analysis of data for the Republic of Ireland, *Social Science and Medicine*, 46, 683–93.

Prothero, R.M. (1965) *Migrants and Malaria*, Longmans, London.

Prothero, R.M. (1977) Disease and mobility: A neglected factor in epidemiology, *International Journal of Epidemiology*, 6, 259–67.

Pyle, G.F. (1969) The diffusion of cholera in the United States in the nineteenth century, *Geographical Analysis*, 1, 59–75.

Raleigh, V.S. and Kiri, V.A. (1997) Life expectancy in England: Variations and trends by gender, health authority, and level of deprivation, *Journal of Epidemiology and Community Health*, 51, 649–58.

Reading, R., Langford, I.H., Haynes, R., and Lovett, A. (1999) Accidents to preschool children: Comparing family and neighbourhood risk factors, *Social Science and Medicine*, 48, 321–30.

Reijneveld, S.A. (1998) Reported health, lifestyles, and use of health care of first generation immigrants in the Netherlands: Do socioeconomic factors explain their adverse position? *Journal of Epidemiology and Community Health*, 52, 298–304.

Remennick, L.I. (1999) Preventive behavior among recent immigrants: Russian-speaking women and cancer screening in Israel, *Social Science and Medicine*, 48, 1669–84.

Richardson, J.L., Langholz, B., Bernstein, L., Burciaga, C., Danley, K., and Ross, R.K. (1992) Stage and delay in breast cancer diagnosis by race, socioeconomic status, age and year, *British Journal of Cancer*, 65, 922–6.

Richardson, S., Monfort, C., Green, M., Draper, G., and Muirhead, C. (1995) Spatial variation of natural radiation and childhood leukaemia incidence in Great Britain, *Statistics in Medicine*, 14, 2487–502.

Rigby, J.E. and Gatrell, A.C. (2000) Spatial patterns in breast cancer incidence in north-west Lancashire, *Area*, 32, 71–8.

Riise, T., Grønning, M., Klauber, M.R., Barrett-Connor, E., Nyland, H., and Albrektsen, G. (1991) Clustering of residence of multiple sclerosis patients at age 13 to 20 years in Hordaland, Norway, *American Journal of Epidemiology*, 133, 932–9.

Roberts, E.M. (1997) Neighborhood social environments and the distribution of low birthweight in Chicago, *American Journal of Public Health*, 87, 597–603.

Robinson, G.M. (1998) *Methods and Techniques in Human Geography*, John Wiley, Chichester.

Rogers, A., ed. (1992) *Elderly Migration and Population Redistribution: A Comparative Study*, Belhaven Press, London.

Rogers, A., Hassell, K., Noyce, P., and Harris, J. (1998) Advice-giving in community pharmacy: Variations between pharmacies in different locations, *Health and Place*, 4, 365–74.

Rogers, D.J. and Williams, B.G. (1993) Monitoring trypanosomiasis in space and time, *Parasitology*, 106, S77–S92.

Rose, G. (1993) *Feminism and Geography: The Limits of Geographical Knowledge*, Polity Press, Cambridge.

Rose, G. (1995) Place and identity: A sense of place, in Massey, D. and Jess, P. (eds.) *A Place in the World?*, Open University Press, Milton Keynes.

Roy, A. (1999) *The Cost of Living*, Flamingo, London.

Sabatier, R. (1996) Migrants and AIDS: Themes of vulnerability and resistance, in Haour-Knipe, M. and Rector, R. (eds.) *Crossing Borders: Migration, Ethnicity and AIDS*, Taylor and Francis, London.

Sabel, C.E., Gatrell, A.C., Löytönen, M., Maasilta, P., and Jokelainen, M. (2000) Modelling exposure opportunities: Estimating relative risk for motor neurone disease in Finland, *Social Science and Medicine*, 50, 1121–37.

Salmond, C., Crampton, P., and Sutton, F. (1998) NZDep91: A New Zealand index of deprivation, *Australian and New Zealand Journal of Public Health*, 22, 95–7.

Salmond, C., Crampton, P., Hales, S., Lewis, S., and Pearce, N. (1999) Asthma prevalence and deprivation: A small area analysis, *Journal of Epidemiology and Community Health*, 53, 476–80.

Saunderson, T. and Langford, I. (1996) A study of the geographical distribution of suicide rates in England and Wales 1989–92 using empirical Bayes estimates, *Social Science and Medicine*, 43, 489–502.

Schærstrom, A. (1996) *Pathogenic Paths? A Time Geographical Approach in Medical Geography*, Lund University Press, Lund.

Schell, L.M. and Czerwinski, S.A. (1998) Environmental health, social inequality and biological differences, in Strickland, S.S. and Shetty, P.S. (eds.) *Human Biology and Social Inequality*, Cambridge University Press, Cambridge.

Schrijvers, C.T.M. and Mackenbach, J.P. (1994) Cancer patient survival by socioeconomic status in the Netherlands: A review for six common cancer sites, *Journal of Epidemiology and Community Health*, 48, 441–6.

Scott, P.A., Temovsky, C.J., Lawrence, K., Gudaitis, E., and Lowell, M.J. (1998) Analysis of Canadian population with potential geographic access to intravenous thrombolysis for acute ischaemic stroke, *Stroke*, 29, 2304–10.

Seale, C. and Pattison, S., eds. (1994) *Medical Knowledge: Doubt and Certainty*, Open University Press, Buckingham.

Senior, M.L. (1991) Deprivation payments to GPs: Not what the doctor ordered, *Environment and Planning C*, 9, 79–94.

Senior, M.L., New, S.J., Gatrell, A.C., and Francis, B.J. (1993) Geographic influences on the uptake of infant immunisations: 2. Disaggregate analyses, *Environment and Planning A*, 25, 467–79.

Senior, M.L., Williams, H.C.W.L., and Higgs, G. (1998) Spatial and temporal variation of mortality and deprivation 2: Statistical modelling, *Environment and Planning A*, 30, 1815–34.

Shannon, G.W. and Dever, G.E. (1974) *Health Care Delivery: Spatial Perspectives*, McGraw-Hill, New York.

Sharp, B.L. and le Sueur, D. (1996) Malaria in South Africa: The past, the present and selected implications for the future, *South African Medical Journal*, 86, 83–9.

Shaw, M., Dorling, D., Gordon, D., and Davey Smith, G. (1999) *The Widening Gap: Health Inequalities and Policy in Britain*, the Policy Press, Bristol.

Shrivistava, P. (1992) *Bhopal: Anatomy of a Crisis*, Paul Chapman, London, second edition.

Slack, R., Ferguson, B., and Ryder, S. (1997) Analysis of hospitalization rates by electoral ward: Relationship to accessibility and deprivation data, *Health Services Management Research*, 10, 24–31.

Slaper, H. Velders, G.J.M., Daniel, J.S. de Gruijl, F.R., and van der Leun, J.C. (1996) Estimates of ozone depletion and skin cancer incidence to examine the Vienna Convention achievements, *Nature*, 384, 256–8.

Smallman-Raynor, M, Muir K.R., and Smith, S.J. (1998) The geographical assignment of cancer units: patient accessibility as an optimal location problem, *Public Health*, 112, 379–83.

Smallman-Raynor, M. and Phillips, D. (1999) Late stages of epidemiological transition: Health status in the developed world, *Health and Place*, 5, 209–22.

Snow, R.W., Craig, M., Deichmann, U., and Marsh, K. (1999) Estimating mortality, morbidity and disability due to malaria among Africa's non-pregnant population, *Bulletin of the World Health Organization*, 77, 624–40.

Sooman, A. and Macintyre, S. (1995) Health and perceptions of the local environment in socially contrasting neighbourhoods in Glasgow, *Health and Place*, 1, 15–26.

Sorlie, P.D. et al. (1995) US mortality by economic, demographic, and social characteristics: the National Longitudinal Mortality Study, *American Journal of Public Health*, 85, 949–56.

Sparks, G., Craven, M.A., and Worth, C. (1994) Understanding differences between high and low childhood accident rate areas: The importance of qualitative data, *Journal of Public Health Medicine*, 17, 193–99.

Stainton Rogers, W. (1991) *Explaining Health and Illness*, Harvester Wheatsheaf, Hemel Hempstead.

Stein, C.E., Fall, C.H., Kumaran, K., Osmond, C., Cox, V., and Barker, D.J. (1996) Fetal growth and coronary heart disease in South India, the *Lancet*, 348, 1269–73.

Stiller, C.A. and Boyle, P.J. (1996) Effect of population mixing and socioeconomic status in England and Wales, 1979–85, on lymphoblastic leukaemia in children, *British Medical Journal*, 313, 1297–1300.

Sun, W.Y., Sangweni, B., Butts, G., and Merlino, M. (1998) Comparisons of immunisation accessibility between non-US born and US-born children in New York City, *Public Health*, 112, 405–8.

Sundquist, J. and Johansson, S.E. (1997) Long-term illness among indigenous and foreign-born people in Sweden, *Social Science and Medicine*, 44, 189–98.

Taylor, D.H. and Leese, B. (1998) General practitioner turnover and migration in England 1990–94, *British Journal of General Practitioners*, 48, 1070–2.

Taylor, H.R. (1995) Ocular effects of UV-B exposure, *Documenta Ophthalmologica*, 88, 285–93.

Teppo, L. (1998) Problems and possibilities in the use of cancer data by GIS: Experience in Finland, in Gatrell, A.C. and Löytönen, M. (eds.) *GIS and Health*, Taylor and Francis, London.

Thamer, M., Richard, C., Casebeer, A.W., and Ray, N.F. (1997) Health insurance coverage among foreign-born US residents: The impact of race, ethnicity, and length of residence, *American Journal of Public Health*, 87, 96–102.

Thiel de Bocanegra, H. and Gany, F. (1997) Providing health services to immigrant and refugee populations in New York, in Ugalde, A. and Cardenas, G. (eds.) *Health and Social Services among International Labor Migrants*, CMAS Books, University of Texas Press, Austin, USA.

Thomas, C. (1999) *Female Forms*, Open University Press, London.

Thomas, J.C. and Thomas, K.K. (1999) Things ain't what they ought to be: social forces underlying racial disparities in rates of sexually transmitted diseases in a rural North Carolina county, *Social Science and Medicine*, 49, 1075–84.

Thomas, R.W. (1992) *Geomedical Systems: Intervention and Control*, Routledge, London.

Thomson, M.C., Connor, S.J., Milligan, P.J.M., and Flasse, S.P. (1996) The ecology of malaria – as seen from earth-observation satellites, *Annals of Tropical Medicine and Parasitology*, 90, 243–64.

Timander, L. and McLafferty, S. (1998) Breast cancer in West Islip, NY: a spatial clustering analysis with covariates, *Social Science and Medicine*, 46, 1623–35.

Townsend, J. (1995) The burden of smoking, in Benzeval, M., Judge, K., and Whitehead, M. (eds.) *Tackling Health Inequalities: An Agenda for Action*, King's Fund, London.

Townsend, P., Phillimore, P., and Beattie, A. (1988) *Health and Deprivation: Inequality and the North*, Croom Helm, London.

Tudor Hart, J. (1971) The inverse care law, the *Lancet*, 27 February, 407–12.

Tunnicliffe, W.S., Burge, P.S., and Ayres, J.G. (1994) Effect of domestic concentrations of nitrogen dioxide on airway responses to inhaled allergen in asthmatic patients, the *Lancet*, 344, 1733–6.

Turrell, G. and Mengersen, K. (2000) Socioeconomic status and infant mortality in Australia: A national study of small urban areas, 1985–89, *Social Science and Medicine*, 50, 1209–25.

Turshen, M. (1984) *The Political Ecology of Disease in Tanzania*, Rutgers University Press, New Brunswick, NJ.

Turshen, M. (1999) *Privatizing Health Services in Africa*, Rutgers University Press, New Brunswick, NJ.

Ugalde, A. (1997) Health and health services utilization in Spain among labor immigrants from developing countries, in Ugalde, A. and Cardenas, G. (eds.) *Health and Social Services among International Labor Migrants*, CMAS Books, University of Texas Press, Austin, USA.

UK Photochemical Oxidants Review Group (1993) *Ozone in the United Kingdom; Third Report*, Department of the Environment, London.

Urry, J. (2000) *Sociology Beyond Societies*, Routledge, London.

Valjus, J., Hongisto, M., Verkasalo, P., Jarvinen, P., Heikkila, K., and Koskenvuo, M. (1995) Residential exposure to magnetic fields generated by 100–400kV power lines in Finland, *Bioelectromagnetics*, 16, 365–76.

Verheij, R.A., de Bakker, D.H., and Groenewegen, P.P. (1999) Is there a geography of alternative medical treatment in The Netherlands? *Health and Place*, 5, 83–98.

Verrept, H. and Louckx, F. (1997) Health advocates in Belgian health care, in Ugalde, A. and Cardenas, G. (eds.) *Health and Social Services among International Labor Migrants*, CMAS Books, University of Texas Press, Austin, USA.

Vianna, N.J. and Polan, A.K. (1984) Incidence of low birthweight among Love Canal residents, *Science*, 226, 1217–9.

Von Reichert, C., McBroom, W.H., Reed, F.W., and Wilson, P.B. (1995) Access to health care and travel for birthing: Native American-white differentials in Montana, *Geoforum*, 26, 297–308.

Wakefield, S. and Elliott, S.J. (2000) Environmental risk perception and well-being: Effects of the landfill siting process in two southern Ontario communities, *Social Science and Medicine*, 50, 1139–54.

Walberg, P., McKee, M., Shkolnikov, V., Chenet, L., and Leon, D.A. (1998) Economic change, crime, and mortality crisis in Russia: Regional analysis, *British Medical Journal*, 1998, 317, 312–8.

Wallace, R. (1990) Urban desertification, public health and public order: "Planned shrinkage", violent death, substance abuse and AIDS in the Bronx, *Social Science and Medicine*, 31, 801–13.

Walsh, S.J., Page, P.H., and Gesler, W.M. (1997) Normative models and healthcare planning: Network-based simulations within a geographic information system environment, *Health Services Research*, 32, 243–60.

Walter, S.D., Birnie, S.E., Marrett, L.D., Taylor, S.M., Reynolds, D., Davies, J., Drake, J.J., and Hayes, M. (1994) The geographic variation of cancer incidence in Ontario, *American Journal of Public Health*, 84, 367–76.

Walters, N.M., Zietsman, H.L., and Bhagwandin, N. (1998) The geographical distribution of diagnostic medical and dental X-ray services in south Africa, *South African Medical Journal*, 88, 383–9.

Wang, T.J. and Stafford, R.S. (1998) National patterns and predictors of β-blocker use in patients with coronary artery disease, *Archives of Internal Medicine*, 158, 1901–6.

Warin, M., Baum, F., Kalucy, E., Murray, C., and Veale, B. (2000) The power of place: Space and time in women's and community health centres in South Australia, *Social Science and Medicine*, 50, 1863–75.

Warnes, A.M., King, R., Williams, A.M., and Patterson, G. (1999) The well-being of British expatriate retirees in southern Europe, *Ageing and Society*, 19, 717–40.

Watson, P. (1995) Explaining rising mortality among men in Eastern Europe, *Social Science and Medicine*, 41, 923–34.

Watts, S. (1997) *Epidemics and History: Disease, Power and Imperialism*, Yale University Press, New Haven.

Watts, S., Khallaayoune, K., Bensefia, R., Laamrani, H., and Gryseels, B. (1998) The study of human behavior and *Schistosomiasis* transmission in an irrigated area in Morocco, *Social Science and Medicine*, 46, 755–65.

Wei, M., Valdez, R.A., Mitchell, B.D., Haffner, S.M., Stern, M.P., and Hazuda, H.P. (1996) Migration status, socioeconomic status, and mortality rates in Mexican Americans and non-hispanic whites: The San Antonio heart study, *Annals of Epidemiology*, 6, 307–13.

Wells, B.L. and Horm, J.W. (1992) Stage at diagnosis in breast cancer: Race and socioeconomic factors, *American Journal of Public Health*, 82, 1383–5.

Whincup, P. and Cook, D. (1997) Blood pressure and hypertension, in Kuh, D. and Ben-Shlomo, Y. (eds.) *A Life Course Approach to Chronic Disease Epidemiology*, Oxford University Press, Oxford.

Whitehead, M. (1995) Tackling inequalities: A review of policy initiatives, in Benzeval, M., Judge, K., and Whitehead, M. (eds.) *Tackling Health Inequalities: An Agenda for Action*, King's Fund, London.

Whiteis, D.G. (1997) Unhealthy cities: Corporate medicine, community economic underdevelopment and public health, *International Journal of Health Services*, 27, 227–42.

Whiteis, D.G. (1998) Third world medicine in first world cities: Capital accumulation, uneven development and public health, *Social Science and Medicine*, 47, 795–808.

Whitelegg, J. (1997) *Critical Mass: Transport, Environment and Society in the Twenty-First Century*, Pluto Press, London.

Whitelegg, J., Gatrell, A.C., and Naumann, P. (1995) The association of health and residential traffic densities, *World Transport, Policy and Practice*, 1, 28–30.

Whitley, E., Gunnell, D., Dorling, D., and Davey Smith, G. (1999) Ecological study of social fragmentation, poverty, and suicide, *British Medical Journal*, 319, 1034–7.

Whittaker, A. (1998) Talk about cancer: Environment and health in Oceanpoint, *Health and Place*, 4, 313–26.

Wilkinson, P., Elliott, P., Grundy, C., Shaddick, G., Thakrar, B., Walls, P., and Falconer, S. (1999) Case-control study of hospital admission with asthma in children aged 5–14 years: relation with road traffic in North West London, *Thorax*, 54, 1070–4.

Wilkinson, R.G. (1996) *Unhealthy Societies: The Afflictions of Inequality*, Routledge, London.

Williams, G. and Popay, J. (1994) Researching the people's health: Dilemmas and opportunities for social scientists, in Popay, J. and Williams, G. (eds.) *Researching the People's Health*, Routledge, London.

Wilson, N., Clements, M., Bathgate, M., and Parkinson, S. (1996) Using spatial analysis to improve and protect the public health in New Zealand, Ministry of Health, New Zealand.

Wing, S., Richardson, D., Armstrong, D., and Crawford-Brown, D. (1997) A reevaluation of cancer incidence near the Three Mile Island nuclear plant: The collision of evidence and assumptions, *Environmental Health Perspectives*, 105, 52–7.

Winter, H., Cheng, K.K., Cummins, C., Maric, R., Silcocks, P., and Varghese, C. (1999) Cancer incidence in the south Asian population of England (1990–92), *British Journal of Cancer*, 79, 645–54.

Wisner, B. (1988) GOBI versus PHC? Some dangers of selective primary health care, *Social Science and Medicine*, 26, 963–9.

Wjst, M., Reitmeir, P., Dold, S., Wulff, A., Nicolai, T., von Loeffelholz-Colberg, E.F., and von Mutius, E. (1993) Road traffic and adverse effects on respiratory health in children, *British Medical Journal*, 307, 596–600.

Wood, E., Yip, B., Gataric, N., Montaner, J.S.G., O'Shaughnessy, M.V., Schechter, M.T., and Hogg, R.S. (2000) Determinants of geographic mobility among participants in a population-based HIV/AIDS drug treatment program, *Health and Place*, 6, 33–40.

Woodward, A., Guest, C., Steer, K., Harman, A., Scicchitano, R., Pisaniello, D., Calder, I., and McMichael, A. (1995) Tropospheric ozone: Respiratory effects and Australian air quality goals, *Journal of Epidemiology and Community Health*, 49, 401–7.

Wu, M-M., Kuo, T-L., Hwang, Y-H., and Chen, C-J. (1989) Dose-response relation between arsenic concentration in well water and mortality from cancers and vascular diseases, *American Journal of Epidemiology*, 130, 1123–32.

Wynne, B. (1996) May the sheep safely graze? A reflexive view of the expert-lay knowledge divide, in Lash, S., Szerszynski, B., and Wynne, B. (eds.) *Risk, Environment and Modernity: Towards a New Ecology*, Sage, London.

Yamada, S., Caballero, J., Matsunaga, D.S., Agustin, G., and Magana, M. (1999) Attitudes regarding tuberculosis in immigrants from the Philippines to the United States, *Family Medicine*, 31, 477–82.

Young, R. (1996) The household context for women's health care decisions: Impacts of UK policy changes, *Social Science and Medicine*, 42, 949–63.

INDEX